NUMBER 153

AMERICAN MATHEMATICAL SOCIETY · INCORPORATED 1923 ·
ΑΓΕΩΜΕΤΡΗΤΟΣ ΜΗ ΕΙΣΙΤΩ

Kenneth C. Millett

Piecewise Linear Concordances and Isotopies

MEMOIRS
OF THE AMERICAN
MATHEMATICAL SOCIETY

Memoirs of the American Mathematical Society
Number 153

Kenneth C. Millett

Piecewise Linear Concordances and Isotopies

Published by the
American Mathematical Society
Providence, Rhode Island
1974

AMS (MOS) subject classifications (1970). Primary 5701, 5720, 5730.

Key words and phrases. Concordance, isotopy, piecewise linear manifolds, fibered general position, fibered sunny collapsing, fibered engulfing, homotopy of spaces of embeddings, block homotopy of spaces of embeddings.

Library of Congress Cataloging in Publication Data CIP

Millett, Kenneth C

Piecewise linear concordances and isotopies.

(Memoirs of the American Mathematical Society, no. 153)

Bibliography: p.

1. Piecewise linear topology. 2. Concordances (Topology) 3. Isotopies (Topology) I. Title. II. Series: American Mathematical Society. Memoirs, no. 153.

QA3.A57 no. 153 [QA613.4] 510'.8s [514'.7] 74-18328

ISBN 0-8218-1853-8

TO DEBORAH

TABLE OF CONTENTS

ABSTRACT

The geometrical techniques in piecewise linear topology are extended to the situation where one has a projection which must be preserved. These include general position, sunny collapsing and engulfing. They are employed to study the relative homotopy groups of the simplicial spaces of concordances and isotopies and, consequently, give a proof of a parameterized version of Hudson's concordance implies isotopy theorem.

Received by the editors March 25, 1974

PIECEWISE LINEAR CONCORDANCES AND ISOTOPIES

INTRODUCTION

A fundamental problem in topology dating from its conception in the works of Hausdorff [34] and Frechet [27], among others, is the description of the local and global structure of various function spaces. The appropriate topology for such spaces has, itself, been the subject of much work, e.g. [8,15,28, 46,48,50,61,62]. These problems still directly provide the impetus for current research. Indeed particular recent interest centers upon various spaces of equivalences, in some category. This paper concerns the structure of the (simplicial) space of piecewise linear homeomorphisms of a piecewise linear manifold.

Until recently very little was known (even locally) of the structure of many of the spaces of homeomorphisms. Low dimensional information is now available through the result of many efforts, e.g. [2,17,25,27,33,50,67,68, 69]. Local information has also become available through the work of others, [18,23,24,49,61]. The situation with regard to global information is similar. With the exception of the classical groups and the researches of Mason [51] and Geoghegan [29], results concern a study of the arc components, e.g., [7, 13,14,20,26], or stable information, e.g. [11,19,30,32,42,46,49,59,64,70,72].

As these results are often achieved only through extraordinary effort, significantly new techniques, or by making substantial restrictions and since, for many purposes, the resulting information has had such profound consequences, e.g., [42,46], or alternative approaches have been developed, e.g., [41,57,64], one might regard the questions concerning the "unstable" range as either insurmountable or uninteresting were it not for the fact that not all manifolds are spheres, i.e., as highly connected, and not all global structure is reflected in the arc components of an individual space. Thus the research begun here is only one step in the attempt to provide the means required to determine the homotopy groups of the (simplicial) piecewise linear homeomorphism group of a piecewise linear manifold although the applications indicated in this paper concern only spheres or Euclidean spaces.

C. Morlet [59] has also begun such a program employing a rather different approach via his "Lemme de disjonction". Also Akiba [2] and Scott [68] have considered the low dimensional problems. The latter approach has its roots in the work of Kuiper and Lashof [44] which also provides the foundation for the point of view employed here.

The basic technique is to relate (or reduce) questions concerning the homotopy groups of homeomorphism spaces to questions concerning homotopy groups of various embedding spaces and, subsequently, to questions concerning the arc components of related embedding spaces. The appropriate connection is hinted at in the work of J.F.P. Hudson, "Concordance, Isotopy and Diffeotopy" [39], which provides one model for the methods developed here. The result is that this paper is primarily concerned with the connections between the structure of concordance spaces and isotopy spaces. From another point of view, the concern is the relationship between the "simplicial" embedding space and the "block" embedding space of Rourke and Sanderson [64].

In Section 1 the basic definitions and statements of the main results are given. The proofs of these results are given in Section 2, assuming Theorem 1.8. The techniques required in proof of Theorem 1.8, given in Section 6, are developed in Sections 3, 4 and 5.

Questions of substantial difficulty remain before the central problem can be effectively answered in a completely general setting. These include the ubiquitous low codimension problems as well as problems concerning the arc components of various concordance (or embedding) spaces and the homotopy groups beyond the range of the main theorem. Many of these are indicated in this paper while others will be discussed in a sequel.

I wish to thank the University of Warwick and Princeton University for the atmosphere and support provided during this research and to R. Edwards, R. Lashof and C. Rourke for their stimulating encouragement and suggestions.

SECTION 1. DEFINITIONS AND STATEMENTS OF RESULTS

All of the definitions and results concern either the simplicial category or the piecewise linear category. For definitions and properties of the simplicial category see Curtis [21] or, for a different point of view, Rourke-Sanderson [66]. All simplicial complexes will satisfy the Kan extension condition. In the main, the basic terminologies, definitions, and notations for the piecewise linear category follow those employed by Zeeman [73] and Hudson [38,39]. Many of these concepts are extended to the fibered situation in this paper. Such extensions will be signaled by the word "fibered" preceding the term. The abuse of the term simplicial complex given for both geometric complexes and the semisimplicial complexes should cause little difficulty as the meaning is always clear from the context.

Let M and N denote compact piecewise linear manifolds of dimensions m and n and boundaries ∂M and ∂N, respectively. Let Δ^s denote the standard s dimensional simplex, $R^n = \{(x_1,\ldots,x_n) | x_i \in R\}$, $R_+^n = \{x \in R^n \mid x_n \geq 0\}$, $R_-^n = \{x \in R^n \mid x_n \leq 0\}$, $I = [0,1]$, $rJ = [-r,r]$, $rD^n = rJ \times \cdots \times rJ$, $1D^n = D^n$, $\Sigma^{n-1} = \partial D^n$, $\Sigma_+^{n-1} = \Sigma^{n-1} \cap R_+^n$, $\Sigma_-^{n-1} = \Sigma^{n-1} \cap R_-^n$, and let B^n denote any piecewise linear homeomorph of D^n. A subspace (K,K_o) of a space (L,L_o) is said to be proper if $K \backslash K_o \subset L \backslash L_o$. A piecewise linear map $f\colon (K,K_o) \to (L,L_o)$ is proper if $(f(K), f(K_o))$ is a proper subcomplex of (L,L_o).

1.1 DEFINITION. Given a proper piecewise linear embedding of a proper subcomplex of $(M,\partial M)$ in $(N,\partial N)$, $f\colon (K,K_o) \to (N,\partial N)$, let $E(M,N;f)$ denote the simplicial complex of locally unknotted proper piecewise linear embeddings of $(M,\partial M)$ into $(N,\partial N)$ extending f. That is the simplicial complex whose s simplices are proper piecewise linear embeddings

$$\begin{array}{ccc} F: (M, \partial M) \times \Delta^s & \longrightarrow & (N, \partial N) \times \Delta^s \\ p \downarrow & & p \downarrow \\ 1: \Delta^s & \longrightarrow & \Delta^s \end{array}$$

such that

(i) the diagram is commutative, i.e., F is a fibered embedding,

(ii) $F^{-1}(\partial N \times \Delta^s) = \partial M \times \Delta^s$, i.e., F is proper,

(iii) $F|(K, K_o) \times \Delta^s = f \times 1$, i.e., F extends f,

(iv) for any simplex Δ linearly embedded in Δ^s, $(N \times \Delta, F(M \times \Delta))$ is a locally unknotted manifold pair, i.e., if B^{n+s} is a regular neighborhood of $x \in F(M \times \Delta)$ in $N \times \Delta$ then there is a proper piecewise linear homeomorphism from $(B^{n+s}, B^{n+s} \cap F(M \times \Delta))$ to (D^{n+s}, D^{m+s}).

If f is not specified the complex is denoted by $E(M,N)$.

1.2 DEFINITION. The complex of germs of proper embeddings of $(M, \partial M)$ into $(N, \partial N)$ extending f is denoted by $GE(M,N;f)$. It consists of the simplices of the complexes of proper embeddings of neighborhoods (U, U_o) of (K, K_o) into $(N, \partial N)$ which extend $f|K_o$.

1.3 DEFINITION. The complex of proper concordances of (germs of) proper embeddings extending a proper embedding $f: (K, K_o) \to (N, \partial)$ of a proper subcomplex of $(M, \partial M)$ is denoted by $C(M, N; f)(GC(M, N; f))$. It is the subcomplex of $E(I \times M, I \times N; 1 \times f)(GE(I \times M, I \times N; 1 \times f))$ whose s-simplices F satisfy

(i) $F^{-1}(\{0\} \times N) = \{0\} \times M$ and

(ii) $F^{-1}(\{1\} \times N) = \{1\} \times M$.

1.4 DEFINITION. The complex of (germs of) proper isotopies of proper embeddings extending a proper embedding $f: (K, K_o) \to (N, \partial N)$ of a proper subcomplex of $(M, \partial M)$ is denoted by $I(M,N;f)(GI(M,N;f))$. It is the complex of paths in $E(M,N;f)(GE(M,N;f))$ or, equivalently, the subcomplex of $E(I \times M, I \times N; 1 \times f)(GE(I \times M, I \times N; 1 \times f))$ such that, for any s-simplex F, the diagram

$$\begin{array}{ccc} F\colon I \times M \times \Delta^s & \longrightarrow & I \times N \times \Delta^s \\ p\downarrow & & p\downarrow \\ 1\colon I \times \Delta^s & \longrightarrow & I \times \Delta^s \end{array}$$

is commutative.

Let $C(M,N)$ and $I(M,N)$ denote the complexes when an f is not specified. Let $i\colon I(M,N) \hookrightarrow C(M,N)$ be the inclusion of $I(M,N)$ as a subcomplex of $C(M,N)$. One purpose of this research is to discover conditions under which $I(M,N)$ is a strong deformation retract of $C(M,N)$. In the simplicial category there are several equivalent statements of this property, e.g., i is a homotopy equivalence, $i_*\colon \Pi_s(I(M,N)) \to \Pi_s(C(M,N))$ is an isomorphism for all s, and $\Pi_s(C(M,N), I(M,N)) = 0$ for all s, cf. [21, Theorem 2.20]. Indeed, it is the last of these three which provides the access to the problem.

J. F. P. Hudson, in Concordance, Isotopy and Diffeotopy [39], proved a more general version of the following theorem, cf. Theorem 1.1 [39].

1.5 THEOREM (Hudson). Let $f\colon (K, \partial M) \to (N, \partial N)$ be a proper piecewise linear embedding and F a 0 simplex of $C(M,N;f)$. If $n - m \geq 3$ there is a proper piecewise linear ambient isotopy H_t of $I \times N$, fixed on $(\{0\} \times N) \cup F(I \times K) \cup (I \times \partial N)$ such that $H_1 F = (1, F \mid (\{0\} \times N))$.

Theorem 1.5 implies that $i_*\colon \Pi_o(I(M,N;f)) \to \Pi_o(C(M,N{:}f))$ is onto if $n - m \geq 3$ or, equivalently, that $\Pi_o(C(M,N;f), I(M,N;f)) = 0$. Furthermore

Hudson's Corollary 1.2 can be employed in the same manner to prove that $\Pi_o(C(M,N),\ I(M,N)) = 0$. This will not be the case here as the basic theorem is not sufficiently strong to provide the necessary information for concordances of embeddings of spheres in spheres.

1.6 DEFINITION. Given piecewise linear spaces (K,K_o) and (L,L_o) a proper fibered piecewise linear embedding of (K,K_o) in (L,L_o) (over Δ^s) is a proper piecewise linear embedding,

$$\begin{array}{ccc} F\colon (K,K_o) \times \Delta^s & \longrightarrow & (L,L_o) \times \Delta^s \\ \Big\downarrow p & & \Big\downarrow p \\ 1\colon \Delta^s & \longrightarrow & \Delta^s \end{array}$$

such that the diagram is commutative.

1.7 DEFINITION. A proper fibered piecewise linear ambient isotopy of (K,K_o) (over Δ^s) is a proper piecewise linear ambient isotopy,

$$\begin{array}{ccc} H_t\colon (K,K_o) \times \Delta^s & \longrightarrow & (K,K_o) \times \Delta^s \\ \Big\downarrow p & & \Big\downarrow p \\ \Delta^s & \longrightarrow & \Delta^s \end{array}$$

such that the diagram is commutative.

The following Theorem 1.8 is the basic result of the paper. Its proof is given in §6.

1.8 THEOREM. Let $f\colon (D^m, \Sigma^{m-1}) \to (N, \partial N)$ be a proper piecewise linear embedding. If $\Pi_j(N) = 0$, $j \le k$, $n - m \ge 3$, $s \le n + k - m - 3$ and F is an s simplex of $C(D^m, N; f \mid \Sigma^{m-1})$ such that $F \mid (I \times D^m \times \partial\Delta^s) = 1 \times f \times 1$, i.e., F is a proper fibered piecewise linear embedding of $I \times (D^m, \Sigma^{m-1}) \times \Delta^s$

in $I \times (N, \partial N) \times \Delta^s$ such that $F|(I \times \Sigma^{m-1} \times \Delta^s) \cup (I \times D^m \times \partial\Delta^s) = 1 \times f \times 1$, then there is a proper fibered piecewise linear ambient isotopy of $I \times N$ over Δ^s, H_t, fixed on $(\{0\} \times N \times \Delta^s) \cup (I \times \partial N \times \Delta^s) \cup (I \times N \times \partial\Delta^s)$, such that $H_1 F = (1, F \mid (\{0\} \times D^m \times \Delta^s))$.

A first consequence of this theorem is a partial answer to the question posed above. Its proof as well as those of the other consequences of Theorem 1.8 are given in §2.

1.9 THEOREM. Let $f\colon \Sigma^{m-1} \to \partial N$ be a piecewise linear embedding. If $\Pi_j(N) = 0$, $j \le k$, and $n - m \ge 3$, then $\Pi_s(C(D^m, N; f), I(D^m, N; f)) = 0$ for all $s \le n + k - m - 3$.

1.10 DEFINITION. Let $f\colon (M, \partial M) \to (N, \partial N)$ be a proper piecewise linear embedding. $C_\partial(M, N; f|K)$ and $I_\partial(M, N; f|K)$ are the subcomplexes of $C(M, N; f|\partial M \cup K)$ and $I(M, N; f|\partial M \cup K)$, respectively, whose simplices F satisfy the property that $F|(\{0,1\} \times M \times \Delta^s) = 1 \times f \times 1$. For $C_{\partial_0}(M, N; F|K)$ and $I_{\partial_0}(M, N; f|K)$ is only necessary that $F|(\{0\} \times M \times \Delta^s) = 1 \times f \times 1$.

A second consequence of Theorem 1.8 is the analogue of Theorem 1.9 for these complexes.

1.11 THEOREM. Let $f\colon (D^m, \Sigma^{m-1}) \to (N, \partial N)$ be a proper piecewise linear embedding. If $\Pi_j(N) = 0$, $j \le k$ and $n - m \ge 3$, then

$$\Pi_s(C_\partial(D^m, N; f|\Sigma^{m-1}), \; I_\partial(D^m, N; f|\Sigma^{m-1})) = 0$$

and

$$\Pi_s(C_{\partial_0}(D^m, N; f|\Sigma^{m-1}), \; I_{\partial_0}(D^m, N; f|\Sigma^{m-1})) = 0$$

for all $s \le n + k - m - 3$.

An important first step toward proving a concordance-isotopy theorem for manifolds, in general, is the following consequence of Theorem 1.11. Indeed it is a special case of the general result whose proof indicates the method to be employed.

1.12 THEOREM. Let $f\colon (D^m \times \Sigma^k, \Sigma^{m-1} \times \Sigma^k) \to (D^m \times \Sigma^n, \Sigma^{m-1} \times \Sigma^n)$ be a proper piecewise linear embedding. If $n - k \geq 3$, then

$$\Pi_s(C_\partial(D^m \times \Sigma^k, D^m \times \Sigma^n;f), I_\partial(D^m \times \Sigma^k, D^m \times \Sigma^n;f)) = 0,$$

and

$$\Pi_s(C_{\partial_o}(D^m \times \Sigma^k, D^m \times \Sigma^n;f), I_{\partial_o}(D^m \times \Sigma^k, D^m \times \Sigma^n;f)) = 0$$

if $s \leq 2n - k - 4$.

A simple consequence of the handle-body theory of piecewise linear manifolds will be required, cf. [38]. Briefly, suppose that M is a piecewise linear manifold and $g\colon \partial D^r \times D^{m-r} \to \partial M$ is a piecewise linear embedding. Let $M' = M \cup_g D^r \times D^{m-r}$ is a piecewise linear manifold gotten from M by attaching an r-handle to M. A handle-body decomposition for a manifold M is a sequence of manifolds

$$D^m = M_o \subset M_1 \subset \cdots \subset M_{s+1} = m$$

where M_{i+1} is obtained from M_i by attaching an r_i-handle.

1.13 PROPOSITION. Let M be a connected piecewise linear manifold. Then M has a handle-body decomposition such that $r_i \geq r_{i+1}$ for all i,

(i) if $\partial M = \emptyset$ there is precisely one 0-handle and one m-handle, and (ii) if $\partial M \neq \emptyset$ there is precisely one 0-handle and no m-handles.

1.14 THEOREM. Let M be a manifold with a handle-body structure $D^m = M_o \subset M_1 \subset \cdots \subset M_{q+1} = M$ such that $\ell \le r_i$ for $0 \le i \le q$, and let $f: (M,\partial M) \to (N,\partial N)$ be a proper embedding into a manifold N with $n-m \ge 3$ and $\Pi_i(N) = 0$ for $j \le k$. Then the inclusions

$$\iota_*: \Pi_s(C_{\partial_o}(M_o,N_o;f),\ I_{\partial_o}(M_o,N_o;f)) \to \Pi_s(C_{\partial_o}(M,N;f),\ I_{\partial_o}(M,N;f))$$

and

$$\iota_*: \Pi_s(C_\partial(M_o,N_o;f),\ I_\partial(M_o,N_o;f)) \to \Pi_s(C_\partial(M,N;f),\ I_\partial(M,N;f))$$

are isomorphisms if $s < n-m+\hat{k}+\ell-3$ and an epimorphism if $s = n-m+\hat{k}+\ell-3$ if $\hat{k} = \min(k, n-m-1)$ and $(N,f(M))$ is constructed from a pair $(N_o,f(M_o))$ by adding relative handles following the construction of M.

1.15 COROLLARY. With the assumptions in Theorem 1.13,

$$\Pi_s(C_{\partial_o}(M,N;f),\ I_{\partial_o}(M,N;f)) \simeq \Pi_s(C_\partial(M,N;f),\ I_\partial(M,N;f)) = 0$$

if $s \le n-m+\hat{k}-3$.

1.16 REMARK. Finally it is appropriate to point out that Theorem 1.8 is not true without restriction. Let G_{n+1} denote the simplicial monoid of self homotopy equivalences of Σ^n and let F_n denote the fiber of the evaluation $F_n \hookrightarrow G_{n+1} \xrightarrow{ev} \Sigma^n$. Note that suspension gives an inclusion of G_n as a subcomplex of F_n. The following proposition is typical of the type of calculation which can be made with these techniques and indicates the complexity which one must anticipate in attempting to extend Theorem 1.9.

1.17 PROPOSITION. If $n \ge 3$ there is a homomorphism

$$\Pi_s(E(D^m,\ D^m \times \Sigma^n;\ j)) \longrightarrow \Pi_{s+m}(\Sigma^n) \oplus \Pi_{s+m}(F_n,G_n),$$

where $j(x) = (x, (1,0,\ldots,0))$, which is an isomorphism if $s < 2n-3$ and an epimorphism if $s = 2n-3$.

Since $\Pi_s(I_\partial(D^o,\Sigma^n)) \simeq \Pi_{s+1}(E(D^o,\Sigma^n)) \simeq \Pi_{s+1}(\Sigma^n)$, in general, and $\Pi_s(C_\partial(D^o,\Sigma^n)) \simeq \Pi_s(E(D^1,D^1 \times \Sigma^n; j))$ the proposition shows that $\iota_*: \Pi_*(I_\partial(D^o,\Sigma^n)) \to \Pi_*(C_\partial(D^o,\Sigma^n))$ cannot always be an isomorphism because there is an epimorphism $\Pi_{2n-2}(C_\partial(D^o,\Sigma^n), I_\partial(D^o,\Sigma^n)) \to \Pi_{2n-2}(F_n,G_n)$ which need not be trivial. Thus Theorem 1.9 and consequently 1.8, is not true in general.

SECTION 2. PROOFS OF CONSEQUENCES OF THEOREM 1.8

This section is concerned with giving proofs of the results announced in §1, assuming Theorem 1.8, whose proof is the subject of the remaining sections of the paper.

2.1 PROOF OF THEOREM 1.9. A homotopy class $\alpha \in \Pi_s(C(D^m,N;f), I(D^m,N;f))$ is represented by an s simplex F, i.e., $\alpha = [F]$,

$$\begin{array}{ccc} F: I \times (D^m,\partial D^m) \times \Delta^s & \longrightarrow & I \times (N,\partial N) \times \Delta^s \\ \downarrow & & \downarrow \\ 1: \Delta^s & \longrightarrow & \Delta^s \end{array}$$

such that $F|I \times (D^m,\partial D^m) \times \partial_o \Delta^s$ is an $s-1$ simplex of $I(D^m,N;f)$ and $F|I \times (D^m,\partial D^m) \times \partial_i \Delta^s = 1 \times f \times 1$, $i = 1,\ldots,s$, where $\partial_i \Delta^s$ denotes the ith face of Δ^s.

Although the following argument is not strictly correct for the simplicial category it can easily be modified by taking appropriate quotients so as to achieve the required delicacy, cf. [21,66]. The difficulty is that the definition of homotopy groups in the simplicial category is rather awkward for geometric purposes. The argument presented here reflects the geometric viewpoint for the purposes of exposition.

Consider the restriction

$$\begin{array}{ccc} F_o = F|\{0\} \times (D^m,\partial D^m) \times \Delta^s & \longrightarrow & \{0\} \times (N,\partial N) \times \Delta^s \\ \downarrow & & \downarrow \\ 1: \Delta^s & \longrightarrow & \Delta^s \end{array}$$

a proper fibered embedding such that $F_o|\partial(\{0\} \times D^m \times \Delta^s) = 1 \times f \times 1$.

Hudson's s-isotopy extension theorem gives a proper fibered piecewise linear ambient isotopy

$$\begin{array}{ccc} G_t\colon \{0\} \times (N,\partial N) \times \Delta^s & \longrightarrow & \{0\} \times (N,\partial N) \times \Delta^s \\ \downarrow & & \downarrow \\ 1\colon \Delta^s & \longrightarrow & \Delta^s \end{array}$$

such that $G_t|(\{0\} \times \partial N \times \Delta^s) = 1$ and $G_1 \circ F_o = 1 \times f \times 1$. Let $(1 \times G_1) \circ F = F'$.

Since $\partial_o F' = F'|I \times (D^m,\partial D^m) \times \partial_o \Delta^s$ is a $s-1$ simplex of $I_{\partial_o}(D^m,N;f)$ the diagram

$$\begin{array}{ccc} \partial_o F'\colon I \times (D^m,\partial D^m) \times \partial_o \Delta^s & \longrightarrow & I \times (N,\partial N) \times \partial_o \Delta^s \\ p(1,3)\downarrow & & \downarrow p(1,3) \\ 1\colon I \times \partial_o \Delta^s & \longrightarrow & I \times \partial_o \Delta^s \end{array}$$

is commutative where $p(1,3)$ is the appropriate projection, cf. Definition 3.3. Hudson's s-isotopy extension theorem [37], actually a slight modification thereof, gives a proper fibered piecewise linear ambient isotopy

$$\begin{array}{ccc} G'_t\colon I \times (N,\partial N) \times \partial_o \Delta^s & \longrightarrow & I \times (N,\partial N) \times \partial_o \Delta^s \\ \downarrow & & \downarrow \\ 1\colon I \times \partial_o \Delta^s & \longrightarrow & I \times \partial_o \Delta^s \end{array}$$

such that the diagram is commutative, G'_t is the identity $(\{0\} \times N \times \partial_o \Delta^s) \cup (I \times \partial N \times \partial_o \Delta^s) \cup (I \times N \times \partial(\partial_o \Delta^s))$, and $G'_1 \circ (\partial_o F') = 1 \times f \times 1$. Via G'_t one can then define a proper fibered piecewise linear homeomorphism

$$\begin{array}{ccc} G'\colon I \times (N,\partial N) \times \Delta^s & \longrightarrow & I \times (N,\partial N) \times \Delta^s \\ p(1,3)\downarrow & & \downarrow p(1,3) \\ I \times \Delta^s & \longrightarrow & I \times \Delta^s \end{array}$$

such that the diagram is commutative, $G'|(\{0\} \times N \times \Delta^s) \cup (I \times \partial N \times \partial^s) = 1$, $G'|(I \times N \times \partial_i \Delta^s) = 1$ for $i = 1,2,\ldots,s$, and $G'F|I \times D^m \times \partial\Delta^s = 1 \times f \times 1$.

Thus, by construction,

$$\begin{array}{ccc} G'F': I \times (D^m,\partial D^m) \times \Delta^s & \longrightarrow & I \times (N,\partial N) \times \Delta^s \\ p(3)\downarrow & & \downarrow p(3) \\ 1: \Delta^s & \longrightarrow & \Delta^s \end{array}$$

is a proper fibered piecewise linear embedding of $I \times (D^m,\partial D^m) \times \Delta^s$ in $I \times (N,\partial N) \times \Delta^s$ such that $G'F'|(I \times \partial D^m \times \Delta^s) \cup (I \times D^m \times \partial\Delta^s) = 1 \times f \times 1$. Therefore Theorem 1.8 asserts the existence of a proper fibered piecewise linear ambient isotopy of $I \times N \times \Delta^s$, H_t', fixed on $(\{0\} \times N \times \Delta^s) \cup (I \times \partial N \times \Delta^s) \cup (I \times N \times \partial\Delta^s)$ such that $H_1'G'F' = (1,G'F'|(\{0\} \times D^m \times \Delta^s))$. Let $H_t = (G' \circ (1 \times G_1))^{-1} \circ H_t' \circ (G' \circ (1 \times G_1))$.

First, H_t is a proper fibered piecewise linear ambient isotopy of $I \times (N,\partial N) \times \Delta^s$ which is fixed on $(\{0\} \times N \times \Delta^s) \cup (I \times \partial N \times \Delta^s) \cup (I \times N \times \partial\Delta^s)$ since H_t' is fixed on that set.

Second, $H_1 \circ F$ is an s simplex of $I(D^m,N;f)$ since

$$\begin{aligned} H_1 \circ F &= (1 \times G_1)^{-1} \circ (G')^{-1} H_1 \circ G' \circ (1 \times G_1) \circ F \\ &= (1 \times G_1)^{-1}(G')^{-1}(1,G'F'|(\{0\} \times D^m \times \Delta^s)) \\ &= (1 \times G_1)^{-1}(1,p(2,3) \circ (G')^{-1} \circ (1,G'F'|(\{0\} \times D^m \times \Delta^s))) \\ &= (1,G_1^{-1} \circ p(2,3) \circ (G')^{-1} \circ (1,G'F'|(\{0\} \times D^m \times \Delta^s))), \end{aligned}$$

i.e., $p(1,3)H_1 \circ F = p(1,3)$ so that $H_1 \circ F \in I(D^m,N;f)$. One might note that the crucial point of the argument is that both $(1 \times G_1)$ and G' respect the projection to $I \times \Delta^s$ and therefore take $1 \times f \times 1$ to an isotopy.

Thus H_t provides a homotopy (abusing the simplicial theory) from α

to the trivial element and concludes the proof of Theorem 1.9.

2.2 PROOF OF THEOREM 1.11. The argument for the pair $(C_{\partial_0}(D^m,N),\ I_{\partial_0}(D^m,N))$ is precisely that given for Theorem 1.9 except that the construction of G_t is unnecessary since $F|\{0\} \times D^m \times \Delta^s = 1 \times f \times 1$ by hypothesis.

The situation for the pair $(C_\partial(D^m,N),\ I_\partial(D^m,N))$ is somewhat more complicated and will require a slightly different proof. Let $\mu\colon I \times I \to I \times I$ denote the simplicial map described by the triangulations given in Figure 2.3 and the requirements that $\mu(0,0) = (0,0)$, $\mu(1,0) = (1,0)$, $\mu(0,1) = (0,1)$, $\mu(1,\frac{1}{2}) = (1,1)$ and, $\mu(1,1) = (0,1)$.

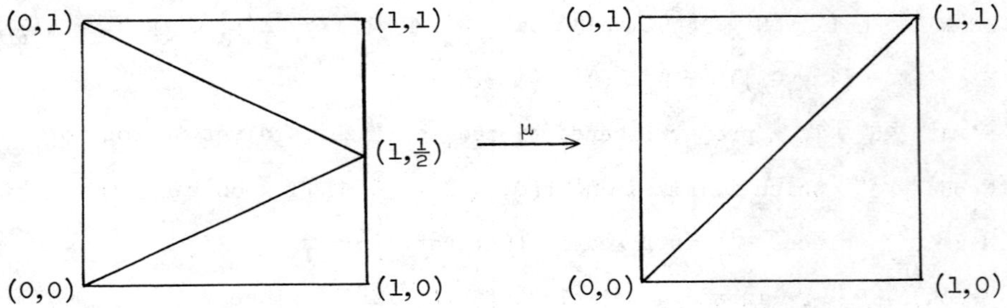

2.3 CONSTRUCTION OF μ.

Similarly, let $\gamma\colon I \times I \to I \times I$ denote the simplicial map described by the triangulations given in Figure 2.4 and the requirements that $\gamma(0,0) = (0,0)$, $\gamma(1,0) = (1,0)$, $\gamma(0,1) = (0,1)$ and, $\gamma(1,1) = (1,\frac{1}{2})$.

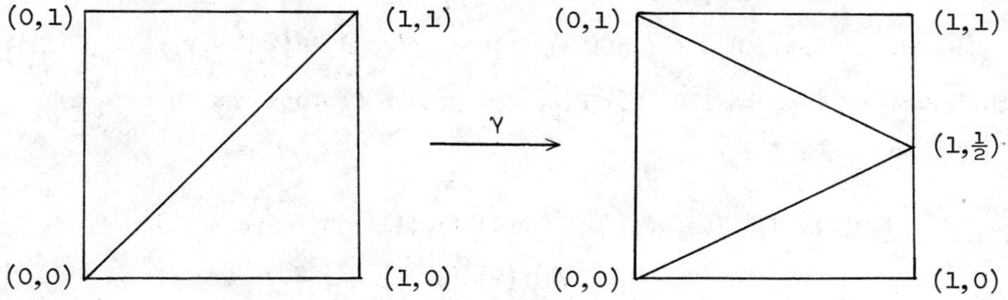

2.4 CONSTRUCTION OF γ.

Let H'_t denote the proper fibered ambient isotopy which is constructed for an s simplex of the pair $(C_\partial(D^m,N), I_\partial(D^m,N))$ considered as an s simplex of the pair $(C_{\partial_o}(D^m,N),\ I_{\partial_o}(D^m,N))$ as above. The required proper fibered ambient isotopy is then constructed from H'_t as follows:

$$H_t(s,x,y) = \begin{cases} (\gamma \circ p(2) \circ \mu(t, p(1) \circ H'_t(s,x,y)), & 0 \le s \le 1 - t/2 \\ \quad p(2,3) \circ H'_t(p(2) \circ \mu(t,s),x,y)) & \\ & \\ (s, p(2,3) \circ H'_{p(1) \circ \mu(t,s)}(1,x,y)), & 1 - t/2 \le s \le 1. \end{cases}$$

First H_t is fixed on $\partial(I \times N \times \Delta^s)$ since

(i) if $s = 0$ $\mu(t,0) = (t,0)$, $H'_t(0,x,y) \circ H'_t(0,x,y) = (0,x,y)$ so that $\gamma \circ p(2)(t,0) = 0$ and $p(2,3)(0,x,y) = (x,y)$,

(ii) if $s = 1$ $\mu(t,1) = (0,1)$, $p(1) \circ \mu(t,1) = 0$ and $p(2,3) \circ H'_o(1,x,y) = (1,x,y)$,

(iii) if $s \in I$, $x \in \partial N$, $y \in \Delta^s$, $H'_{t'}(s',x,y) = (s',x,y)$ for all t', s' so that it is sufficient to observe that $\gamma \circ p(2) \circ \mu(t,s) = s$ if $0 \le s \le 1 - t/2$ and,

(iv) if $s \in I$, $x \in N$, $y \in \partial\Delta^s$, $H'_{t'}(s',x,y) = (s',x,y)$ for all

t', s' so that this case is the same as (iii).

Second, $p(1,3) \circ H_1 \circ F = p(1,3)$ since $p(1,3) \circ H_1' \circ F(s,x,y) = p(1,3)$. By the proof of Theorem 1.9 $H_1' \circ F(s,x,y)$ is an isotopy and therefore

(i) if $0 \le s \le \frac{1}{2}$

$$\begin{aligned} p(1) \circ H_1 \circ F(s,x,y) &= \gamma \circ p(2)\mu(1, p(1) \circ H_1' \circ F(s,x,y)) \\ &= \gamma \circ p(2)\mu(1,s) \\ &= s \end{aligned}$$

and if $\frac{1}{2} \le s \le 1$, $p(1) \circ H_1 \circ F(s,x,y) = s$ and

(ii) if $0 \le s \le \frac{1}{2}$

$$\begin{aligned} p(3) \circ H_1 \circ F(s,x,y) &= p(3) \circ H_1'(p(2) \circ \mu(1, p(1) \circ F(s,x,y)), x, y) \\ &= y \end{aligned}$$

and if $\frac{1}{2} \le s \le 1$

$$\begin{aligned} p(3) \circ H_1 \circ F(s,x,y) &= p(3)H_{t'}'(1, p(2,3) \circ F(s,x,y)) \\ &= p(3)F(s,x,y) = y. \end{aligned}$$

Thus H_t is a homotopy in $C_\partial(D^m, N)$ from F, $[F] = \alpha$, to $H_1F \in I_\partial(D^m, N)$ which is fixed on $I \times N \times \Delta^s$. Therefore $\alpha = 0$ and $\Pi_s(C_\partial(D^m,N), I_\partial(D^m,N)) = 0$.

Several simplicial fibrations will be required in the proofs of Theorems 1.12 and 1.13. These are recognized via the standard methods, cf. [44,59], by employing Hudson's s covering isotopy theorem, Theorem 4.1 of [39].

2.5 PROPOSITION. Let

$$(D^{k-1}, \Sigma^{k-1}) \overset{\iota}{\hookrightarrow} (M, \partial M) \xrightarrow{f} (N, \partial N)$$

be embeddings with $n - k \ge 3$ and $n - m \ge 3$.

(i) $E(M,N;f\iota|D^k) \hookrightarrow E(M,N;f) \to E(D^k,N;f\iota)$ is a simplicial fibration.

(ii) If $D^k = D^i \times D^j$

$$E(M,N;f \mid \partial M \cup \iota(D^i \times \tfrac{1}{2} D^j)) \hookrightarrow$$
$$E(M,N;f \mid \partial M \cup \iota(D^i \times \{0\})) \longrightarrow$$
$$GE(D^i \times D^j,N;f\iota \mid \partial D^i \times D^j \cup D^i \times \{0\})$$

is a simplicial quasi fibration, i.e., the associated homotopy sequence is exact.

2.6 REMARK. Indeed Proposition 2.5 (ii) would be a simplicial fibration if the neighborhood of the germ had not been normalized to $D^i \times \frac{1}{2} D^j$. Furthermore these imply the existence af analogous fibrations for concordance and isotopy spaces merely by employing the definitions 1.3 and 1.4, respectively. It is the pairs of these fibrations that are actually required.

2.7 PROOF OF THEOREM 1.12. The proof is by induction on k. The case $k = 0$ follows from Theorem 1.11 where one has

$$\Pi_s(C_{\partial_o}(D^m,D^m \times \Sigma^n),\ I_{\partial_o}(D^m,D^m \times \Sigma^n)) = 0$$

if $s \le (m+n) - (n-1) - m - 3 = 2n - 4$. To complete this case, employ the fibration

$$E(D^m \times \Sigma^o, D^m \times \Sigma^n;f| \Sigma^{n-1} \times \Sigma^o \cup D^m \times \{1\}) \to E(D^m \times \Sigma^o, D^m \times \Sigma^n;f| \Sigma^{m-1} \times \Sigma^o)$$
$$\to E(D^m \times \{0\},D^m \times \Sigma^n;f| \Sigma^{m-1} \times \{1\})$$

to show that, since $E(D^m \times \Sigma^o, D^m \times \Sigma^n;f| \Sigma^{m-1} \times \Sigma^o \cup D^m \times \{1\})$ is contractible, the projection is a homotopy equivalence. From the definitions of C_{∂_o} and I_{∂_o} this implies that

$$\Pi_s(C_{\partial_o}(D^m \times \Sigma^o, D^m \times \Sigma^n), I_{\partial_o}(D^m \times \Sigma^o, D^m \times \Sigma^n))$$
$$\simeq \Pi_s(C_{\partial_o}(D^m, D^m \times \Sigma^n), I_{\partial_o}(D^m, D^m \times \Sigma^n)) = 0$$

for $s \le 2n - 4$.

Suppose the theorem has been proved for $0,1,2,\ldots,k-1$. Consider the homotopy sequence of the pair of fibrations

$$\begin{array}{c}(C_{\partial_o}(D^m \times \Sigma^k, D^m \times \Sigma^n; f|D^m \times \{1\}), I_{\partial_o}(D^m \times \Sigma^k, D^m \times \Sigma^n; f|D^m \times \{1\}))\\ \downarrow \\ (C_{\partial_o}(D^m \times \Sigma^k, D^m \times \Sigma^n), I_{\partial_o}(D^m \times \Sigma^k, D^m \times \Sigma^n))\\ \downarrow \\ (C_{\partial_o}(D^m \times \{0\}, D^m \times \Sigma^n), I_{\partial_o}(D^m \times \{0\}, D^m \times \Sigma^n)).\end{array} \tag{2.8}$$

As observed previously the homotopy groups of the base are zero when $s \le 2n - 4$. To analyze the homotopy groups of the fiber consider the homotopy sequences of the pairs of fibrations

$$\begin{array}{c}(C_{\partial_o}(D^m \times \Sigma^k, D^m \times \Sigma^n; f|D^m \times \Sigma^k_+), I_{\partial_o}(D^m \times \Sigma^k, D^m \times \Sigma^n, f|D^m \times \Sigma^k_+)\\ \downarrow \\ (C_{\partial_o}(D^m \times \Sigma^k, D^m \times \Sigma^n; f|D^m \times \{1\}), I_{\partial_o}(D^m \times \Sigma^k, D^m \times \Sigma^n; f|D^m \times \{1\}))\\ \downarrow \\ (GC_{\partial_o}(D^m \times \Sigma^k, D^m \times \Sigma^n; f|D^m \times \{1\}), GI_{\partial_o}(D^m \times \Sigma^k, D^m \times \Sigma^n; f|D^m \times \{1\}))\end{array} \tag{2.9}$$

and (2.10)

$$\begin{array}{c}(C_{\partial_o}(D^m \times D^k, D^m \times D^n; f|D^m \times (\Sigma^{k-1} \cup \tfrac{1}{2} D^k)), I_{\partial_o}(D^m \times D^k, D^m \times D^n; f|D^m \times (\Sigma^{k-1} \cup \tfrac{1}{2} D^k)))\\ \downarrow \\ (C_{\partial_o}(D^m \times D^k, D^m \times D^n; f|D^m \times (\Sigma^{k-1} \cup \{0\})), I_{\partial_o}(D^m \times D^k, D^m \times D^n; f|D^m \times (\Sigma^{k-1} \times \{0\})))\\ \downarrow \\ (GC_{\partial_o}(D^m \times D^k, D^m \times D^n; f|D^m \times \{0\}), GI_{\partial_o}(D^m \times D^k, D^m \times D^n; f|D^m \times \{0\}))).\end{array}$$

First note that the base pairs are equivalent since the germs do not, in fact, depend upon the ambient manifold. Next observe that the fiber of (2.9) and the total space of (2.10) are contractible via the usual generalization of the Alexander Isotopy, cf. [44]. All that remains is to apply the induction hypothesis to the fiber of (2.10) which is equal to

$$(C_{\partial_o}(D^{m+1} \times \Sigma^{k-1}, D^{m+1} \times \Sigma^{n-1}), \ I_{\partial_o}(D^{m+1} \times \Sigma^{k-1}, D^{m+1} \times \Sigma^{n-1})).$$

By induction, the sth homotopy group of this pair is zero if $s \leq 2(n-1) - (k-1) - 4 = 2n - k - 5$ and therefore the exact sequences imply that

$$\Pi_s(C_{\partial_o}(D^m \times \Sigma^k, D^m \times \Sigma^n; f), \ I_{\partial_o}(D^m \times \Sigma^k, D^m \times \Sigma^n; f)) = 0$$

if $s \leq 2n - k - 4$.

An analogous argument proves this result for

$$\Pi_s(C_\partial(D^m \times \Sigma^k, D^m \times \Sigma^k; f), \ I_\partial(D^m \times \Sigma^k, D^m \times \Sigma^k; f)).$$

The proof of Theorem 1.13 will be given by an analysis of a handle-body structure for the pair $(N, f(M))$. The inductive case, induction on the number of handles, is described in the following proposition.

2.11 PROPOSITION. Suppose that $f_o: (M_o, \partial M_o) \to (N_o, \partial N_o)$ is a proper embedding with $n - m \geq 3$. Furthermore suppose that $f_1: (M_1, \partial M_1) \to (N_1, \partial N_1)$ is a proper embedding obtained from f_o by adding a p handle to the pair $(N_o, f(M_o))$ and $\Pi_j(N_1) = 0$ for $j \leq k$. Then the natural inclusion induced homomorphism from

$$\iota_*: \Pi_s(C_{\partial_o}(M_o, N_o; f_o), \ I_{\partial_o}(M_o, N_o; f_o)) \to \Pi_s(C_{\partial_o}(M_1, N_1; f_1), \ I_{\partial_o}(M_1, N_1; f_1))$$

is an isomorphism if $s < n - m + \hat{k} + p - 3$ and an epimorphism if $s = n - m + \hat{k} + p - 3$ where $\hat{k} = \min(k, n - m - 1)$.

2.12 PROOF OF PROPOSITION 2.11. Let $i: (D^{m-p}, \Sigma^{m-p-1}) \to (M_1, \partial M_1)$ denote the inclusion of the cell transverse to the handle in M_1. Proposition 2.5 implies the existence of a pair of fibrations

$$\text{(2.13)}\qquad \begin{array}{ccc} I_{\partial_o}(M_1,N_1;f_1|i(D^{m-p})) & \to & C_{\partial_o}(M_1,N_1;f_1|i(D^{m-p})) \\ \downarrow & & \downarrow \\ I_{\partial_o}(M_1,N_1;f_1) & \longrightarrow & C_{\partial_o}(M_1,N_1;f_1) \\ \downarrow & & \downarrow \\ I_{\partial_o}(D^{m-p},N_1;f_1) & \longrightarrow & C_{\partial_o}(D^{m-p},N_1;f_1). \end{array}$$

Theorem 1.11 implies that the homomorphism relating the homotopy groups of the base spaces is an isomorphism if $s < n+k-(m-p)-3$ and an isomorphism if $s = n+k-(m-p)-3$.

The fibers of (2.13) are studied via the fibration (2.14) where $j: \frac{1}{2}D^p \times D^{m-p} \to M_1$ is the inclusion of a product neighborhood of the transverse cell

$$\text{(2.14)}\qquad \begin{array}{ccc} I_{\partial_o}(M_1,N_1;f_1|j(\frac{1}{2}D^p \times D^{m-p})) & \to & C_{\partial_o}(M_1,N_1;f_1|j(\frac{1}{2}D^p \times D^{m-p})) \\ \downarrow & & \downarrow \\ I_{\partial_o}(M_1,N_1;f_1|i(D^{m-p})) & \longrightarrow & C_{\partial_o}(M_1,N_1;f_1|i(D^{m-p})) \\ \downarrow & & \downarrow \\ GI_{\partial_o}(M_1,N_1;f_1|i(D^{m-p})) & \longrightarrow & GC_{\partial_o}(M_1,N_1;f_1|i(D^{m-p})). \end{array}$$

To accomplish this first note that the base spaces depend neither upon N_1 nor f_1 and are equivalent to the base spaces of the following fibrations.

$$\begin{array}{ccc}
I_{\partial_o}(D^p \times D^{m-p}, D^n; i|(\tfrac{1}{2}D^p \times D^{m-p})) & \rightarrow & C_{\partial_o}(D^p \times D^{m-p}, D^n; i|(\tfrac{1}{2}D^p \times D^{m-p})) \\
\downarrow & & \downarrow \\
I_{\partial_o}(D^p \times D^{m-p}, D^n; i(D^{m-p})) & \longrightarrow & C_{\partial_o}(D^p \times D^{m-p}, D^n; i(D^{m-p})) \\
\downarrow & & \downarrow \\
GI_{\partial_o}(D^p \times D^{m-p}, D^n; i(D^{m-p})) & \longrightarrow & GC_{\partial_o}(D^p \times D^{m-p}, D^n; i(D^{m-p})).
\end{array} \tag{2.15}$$

The two total spaces of (2.15) are contractible via the standard generalization of the Alexander isotopy, [44], while the fibers are easily recognized as $I_{\partial_o}(D^{m-p+1} \times \Sigma^{p-1}, D^{m-p+1} \times \Sigma^{n-m+p-1})$ and $C_{\partial_o}(D^{m-p+1} \times \Sigma^{p-1}, D^{m-p+1} \times \Sigma^{n-m+p-1})$, respectively. Theorem 1.12 implies that the inclusion is an isomorphism on homotopy groups for $s < 2(n-m+p-1) - (p-1) - 4 = 2(n-m) + p - 5$ and an epimorphism for $s = 2(n-m) + p - 5$. Consequently the homomorphism on the base space of (2.15), and hence (2.14), is an isomorphism for $s < 2(n-m) + p - 4$ and an epimorphism for $s = 2(n-m) + p - 4$.

The fibers of (2.14) are similarly recognized as $I_{\partial_o}(M_o, N_o; f_o)$ and $C_{\partial_o}(M_o, N_o; f_o)$, respectively. The exact sequence of the pair of fibrations in (2.14) and the previous remark give that

$$\Pi_s(C_{\partial_o}(M_o, N_o; f_o), I_{\partial_o}(M_o, N_o; f_o)) \longrightarrow \Pi_s(C_{\partial_o}(M_1, N_1; f_1 | i(D^{m-p})), I_{\partial_o}(M_1, N_1; f_1 | i(D^{m-p})))$$

is an isomorphism if $s < 2(n-m) + p - 4$ and an epimorphism if $s = 2(n-m) + p - 4$.

Combining this with the previous information for (2.13) we have an isomorphism if $s < n + k - (m-p) - 3$ and $s < 2(n-m) + p - 4$. This can be rewritten as $s < n - m + \hat{k} + p - 3$ where $\hat{k} = \min(k, n-m-1)$.

2.16 PROOF OF THEOREM 1.13. By induction on the number of handles and the observation that the associated N_i's are $\hat{k}$ connected the previous proposition proves that

$$\Pi_s(C_{\partial_o}(M_o,N_o;f_o),\ I_{\partial_o}(M_o,N_o;f_o)) \longrightarrow \Pi_s(C_{\partial_o}(M,N;f),\ I_{\partial_o}(M,N;f))$$

is an isomorphism if $s < n - m + \hat{k} + \ell - 3$ and an epimorphism if $s = n - m + \hat{k} + \ell - 3$.

Similarly for $\Pi_s(C_\partial(M,N;f),\ I_\partial(M,N;f))$.

2.17 PROOF OF COROLLARY 1.14. By Theorem 1.11 $\Pi_s(C_{\partial_o}(M_o,N_o;f_o),\ I_\partial(M_o,N_o;f_o)) = 0$ if $s \le n + \hat{k} - m - 3$ so that, by Theorem 1.13, $\Pi_s(C_{\partial_o}(M,N;f),\ I_{\partial_o}(M,N;f) = 0$ for $s \le n + \hat{k} - m - 3$.

2.18 PROOF OF PROPOSITION 1.17. The proof is by induction on s. The case $s = 0$ is precisely Corollary 2.18, page 454, of Rourke and Sanderson [64] which states in effect that $\Pi_o(E(D^m, D^m \times \Sigma^n;j))$ is isomorphic to $\Pi_m(G_{n+1},G_n)$ for $n > 2$. It is easily noted that $\Pi_m(G_{n+1},G_n) \cong \Pi_m(\Sigma^n) \oplus \Pi_m(F_n,G_n)$ and that the homomorphism from $\Pi_o(E(D^m,D^m;j))$ onto the $\Pi_m(\Sigma^n)$ summand is defined by taking the homotopy class of the projection of the embedding, relative to its boundary, in Σ^n.

Assume, inductively, that the proposition has been verified for $s = 0,1,\ldots,k-1$. First $\Pi_k(E(D^m,D^m \times \Sigma^n;j)) \cong \Pi_{k-1}(\Omega E(D^m,D^m \times \Sigma^n;j))$. Since $\Omega E(D^m,D^m \times \Sigma^n;j) = I_\partial(D^m,D^m \times \Sigma^n)$, $\Pi_{k-1}(\Omega E(D^m,D^m \times \Sigma^n;j)) \cong \Pi_{k-1}(I_\partial(D^m,D^m \times \Sigma^n))$. By Theorem 1.11, if $n \ge 3$ and $k < 2n-3$ $(k = 2n-3)$, the inclusion of $I_\partial(D^m,D^m \times \Sigma^n)$ in $C_\partial(D^m,D^m \times \Sigma^n)$ induces an isomorphism (epimorphism) on Π_{k-1}. Furthermore $C_\partial(D^m,D^m \times \Sigma^n)$ is $E(D^{m+1},D^{m+1} \times \Sigma^n;j)$ so that by induction $\Pi_{k-1}(C_\partial(D^m,D^m \times \Sigma^n))$ is isomorphic to $\Pi_{k+m}(\Sigma^n) \oplus \Pi_{k+m}(F_n,G_n)$. The composition of these isomorphisms (epimorphisms) gives the isomorphism (epimorphism)

$$\Pi_k(E(D^m, D^m \times \Sigma^n; j)) \longrightarrow \Pi_{k+m}(\Sigma^n) \oplus \Pi_{k+m}(F_n, G_n).$$

One also notes that the projection of $\Pi_k(E(D^m, D^m \times \Sigma^n; j))$ onto the $\Pi_{k+m}(\Sigma^n)$ summand is defined by taking the homotopy class of the projection of the k simplex F of embeddings, relative to the total boundary, $p(2) \circ F\colon D^m \times \Delta^k \to \Sigma^n$.

SECTION 3. FIBERED GENERAL POSITION

The two general position techniques described in this section are substantial improvements upon previously described methods for fibered situations [32, 49, 52]. The first is sufficient for applications involving Lemma 4.6 while the second, the more complicated of the two, is employed with Lemma 4.16 to provide a portion of the proof of the main result of the paper, i.e., it is used to prove Theorem 5.1. Before stating these lemmas several definitions are required.

3.1 DEFINITION. Let K and L be simplicial complexes and $p: K \to L$ a (specified) simplicial map. If α is a simplex of K the projective dimension of α, $\text{pdim}(\alpha)$, is $\dim p(\alpha)$. The fiber dimension of α, $\text{fpdim}(\alpha)$, is $\dim(\alpha) - \text{pdim}(\alpha)$. The fiber codimension of α, $\text{fcpdim}(\alpha)$, is the maximum of $\dim(\beta) - \dim(\alpha)$, where α is a face of β and $p(\alpha) = p(\beta)$.

3.2 DEFINITION. Let K and L be simplicial complexes and $f: K \to L$ a simplicial map. The singularity set of f,

$$S(f) = \{x \in K \mid f^{-1}(f(x)) \neq \{x\}\}.$$

Note that $S(f)$ is not a subcomplex but its closure is a subcomplex. f is said to be nonsingular if $S(f) = \emptyset$ and nondegenerate if for every $\sigma \in K$, $S(f|\sigma) = \emptyset$.

3.3 DEFINITION. The function $p(B): \Pi_{\alpha \in A} X_\alpha \to \Pi_{\beta \in B} X_\beta$, $B \subset A$, is the usual projection.

3.4 LEMMA. Let $(N,\partial N)$ be a compact piecewise linear manifold and (K,K_o) a subcomplex of $I \times (N,\partial N) \times \Delta^s$. Given $\epsilon > 0$ there are triangulations T_1 and T_2 of $I \times (N,\partial N) \times \Delta^s$ and a proper fibered piecewise linear ϵ ambient isotopy, φ_t, of $I \times (N,\partial N) \times \Delta^s$ such that φ_1 is simplicial from T_1 to T_2 and commuting with $p(1)$, i.e., "level preserving", with the following properties: if σ and τ are simplices of $\varphi_1(K)$ not in $\varphi_1(K_o)$ $(\varphi_1(K_o))$ with respect to T_2, then if $p = p(2,3)$ and $p' = p(3)$

(i) $\dim(p(\sigma) \cap p(\tau)) \le p\dim(\sigma) + p\dim(\tau) - \max(p'\dim(\sigma), p'\dim(\tau)) - \dim(N)$ $(\dim(\partial N))$,

(ii) $fp'\dim(p(\sigma) \cap p(\tau)) \le \max(0, fp'\dim(p(\sigma)) + fp'\dim(p(\tau)) - \dim(N)$ $(\dim(\partial N)))$,

(iii) if $fp'\dim(K)$ $(fp'\dim(K_o)) \le \dim(N)$ $(\dim(\partial N))$ then $p \circ \varphi_1|K$ $(p \circ \varphi_1|K_o)$ is nondegenerate.

Furthermore, if (i), (ii) and (iii) are satisfied for the simplices of (K',K'_o) a subcomplex of (K,K_o) and $(K,K_o) \cap (I \times (L,L_o) \times \Delta^s) \subset (K',K'_o)$, (L,L_o) a subcomplex of $(N,\partial N)$ then φ_t may be taken to be a proper fibered piecewise linear ϵ ambient isotopy of $I \times (N,\partial N) \times \Delta^s$ relative to $(K',K'_o) \cup (I \times (L,L_o) \times \Delta^s)$.

3.5 REMARK. With the exception of condition (ii) this is a standard fibered general position result. Condition (ii) is a very crucial improvement in that it requires that the fiber dimension of the singularity set of $p \circ \varphi_1|K$ be as small as possible. This improvement is illustrated by a comparison of figures 3.6 and 3.7. In Figure 3.6 the fiber dimension of $\sigma \cap \tau$ is 1. In Figure 3.7, where the vertex v has been moved within the

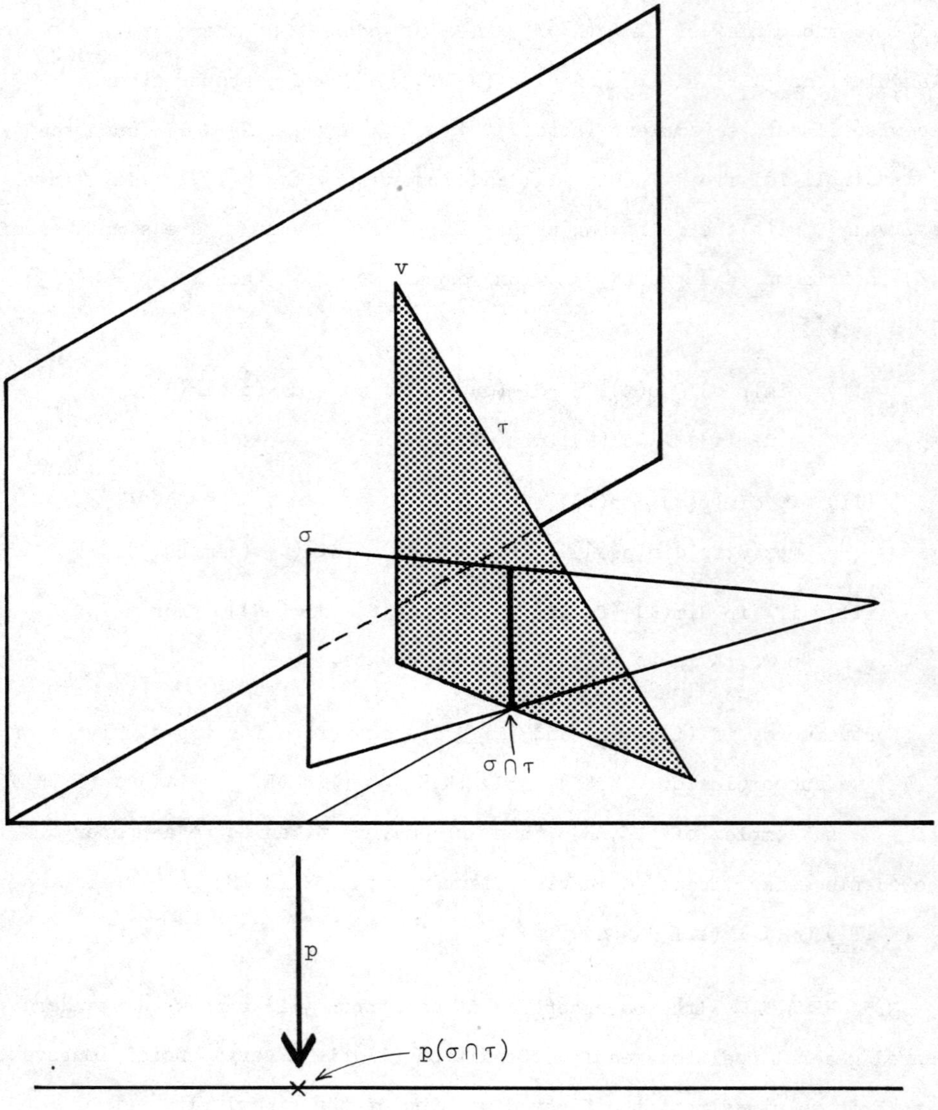

3.6 THE RESULT OF USUAL FIBERED GENERAL POSITION.

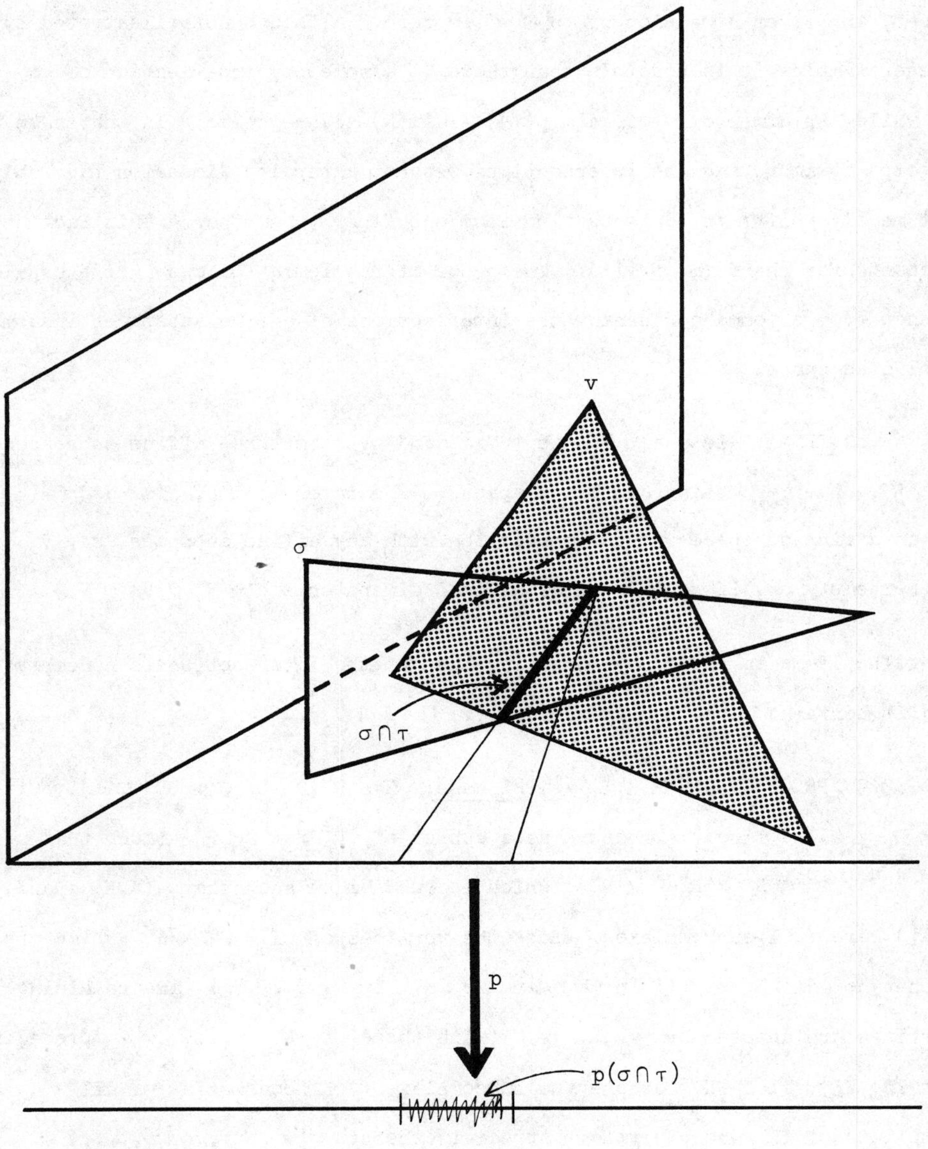

3.7 THE RESULT OF NEW FIBERED GENERAL POSITION

fiber, the fiber dimension of $\sigma \cap \tau$ is zero. Although not illustrated by these examples it is possible that there be a nonempty intersection of σ and τ while $fp'\dim(p(\sigma)) + fp'\dim(p(\tau)) - \dim(N)$ is negative. In that case any attempt at improving the intersection does not alter its dimension favorably but merely shifts the intersection from one fiber to another. This fact accounts for the form of (ii). The proof of the lemma follows in the usual manner from a lemma concerning the intersections of affine subspaces in some Euclidean space.

3.8 LEMMA. Let $a + V_1$, $b + V_2$, and V_3 be three affine subspaces of R^n, the V_i being subvector spaces. The intersection, $(a+V_1) \cap (b+V_2)$ is an affine subspace whose intersection with any affine subspace $c + V_3$ is either empty or, if nonempty, has a fixed dimension, $\dim(V_1 \cap V_2 \cap V_3)$.

The lemma is proved by observing that if the intersection is nonempty, say it contains d, then it is $d + (V_1 \cap V_2 \cap V_3)$.

3.9 PROOF OF LEMMA 3.4. <u>First case:</u> $N = D^n$, $\partial D^n \subset L(\partial D^n_- \subset L)$. Give $I \times D^n \times \Delta^s$ a linear structure as a subset of $R^1 \times R^n \times R^s$ (such that $I \times \partial D^n_+ \times \Delta^s \subset R^1 \times R^{n-1} \times R^s$) which is subdivided such that (K,K_o) and (K',K'_o) are full subcomplexes. List the vertices of $I \times D^n \times \Delta^s$ which are either not in K or, if in K, are in K' by $v_1,\ldots,v_k$. The remaining vertices are denoted by $w_1,\ldots,w_\ell$ (with those of K_o $w_1,\ldots,w_{\ell'}$ preceding not in K_o). Let K_i denote the subcomplex of K consisting of all simplices of K whose vertices appear in the set $\{v_1,\ldots,v_k,w_1,\ldots,w_i\}$. By construction K_i is a full subcomplex of K_j, $i < j$, and $K = K_\ell$.

The proof of this case is by induction where, in the statement of the lemma, K is replaced by K_{i+1}, K' by K_i and ε by ε/ℓ. Consider the simplices of $K_{i+1} \setminus K_i$ and the proper affine subspaces which they determine in $R^1 \times R^n \times R^s$ (in $R^1 \times R^{n-1} \times R^s$ if $i + 1 \leq \ell'$). The intersections of these spaces with the fiber of $p(1,3)$ containing w_{i+1} and those which are parallel and contain w_{i+1} form a set, the union of the subspaces, which has measure zero and hence a complement which is dense in the fiber. Let w'_{i+1} be a point in this complement, within the star of w_{i+1} ($R^1 \times R^{n-1} \times R^s$ if $i + 1 \leq \ell'$), and within ϵ/ℓ of w_{i+1}.

The linear translation of w_{i+1} to w'_{i+1} extends to the desired piecewise linear fibered ϵ/ℓ ambient isotopy relative to $K' \cup I \times L \times \Delta^s$. It is necessary to verify properties (i), (ii) and (iii) for the given triangulation. This will not imply them for further subdivisions!

(i) To compute the dimension of $p(\sigma) \cap p(\tau)$ first assume that $fp'\dim(\sigma)$ is smaller than $fp'\dim(\tau)$. Furthermore, as the worst possible case, take $p(\sigma)$ to lie entirely in a single fiber, i.e., $p'(\sigma)$ is a single vertex. Under these assumptions $\dim(p(\sigma) \cap p(\tau)) \leq fp'\dim(p(\sigma)) + p\dim(\tau) - \dim(N)$ $(\dim(\partial N))$. The statement in (i) expresses this fact independently of the assumption on $fp'\dim(\sigma)$ since $fp'\dim(\alpha) = \dim(\alpha) - p'\dim(\alpha)$ and $fp'\dim(\alpha) = fp'\dim(p(\alpha))$.

(ii) Assume that $\dim(p(\sigma) \cap p(\tau)) \geq 0$. Thus the intersection is nonempty and the dimension of its intersection with a fiber, if nonempty, is constant by the lemma. This dimension is, at most, $fp'\dim(p(\sigma)) + fp'\dim(p(\tau)) - \dim(N)$ $(\dim(\partial N))$ since the affine subspaces are in general position in each fiber.

<u>Second case</u>: The general case. Let $I \times (N, \partial N) \times \Delta^s$ be triangulated so that $p(2,3)$ and $p(2)$ are simplicial and so that there is a collection of simplicial subballs of N, $\{B_{i,j}\}$ $i \leq i \leq m$, $j = 1,2$, whose interiors

cover N and such that $B_{i,2} \subset \text{int } B_{i,1}$ and $B_{i,j} \cap \partial N$ is either empty or $\partial(B_{i,j})_+$.

The proof is by induction where, in the statement of the lemma, (K,K_o) is replaced by $(K,K_o) \cap (I \times (B_{i,j} \cup \bigcup_{j=1}^{i-1} B_{j,2}) \times \Delta^s)$, (K',K_o') by $((K,K_o) \cap (I \times (\bigcup_{j=1}^{i-1} B_{j,2}) \times \Delta^s)) \cup (K',K'_o)$ and ϵ by ϵ/m. This case follows from the first when N is replaced by $B_{i,1}$, (K,K_o) by $((K,K_o) \cap B_{i,2}) \cup (K,K_o) \cap B_{i,j} \cap (\bigcup_{j=1}^{i-1} B_{j,2}))$, (K',K_o') by $((K',K_o') \cap B_{i,j}) \cup ((K,K_o) \cap B_{i,1} \cap (\bigcup_{j=1}^{i-1} B_{j,2}))$, and L by $(L \cap B_{i,1}) \cup (\partial B_{i,1})$, or $(L \cap B_{i,1}) \cup (\partial(B_{i,1})_-)$ if $\partial N \cap B_{i,1} = \partial(B_{i,1})_+$. The ambient isotopy of the first case extends to the identity on the complement of $B_{i,1}$ and gives the inductive conclusion required to complete the proof of the lemma.

3.10 REMARK. The following lemma 3.12 makes full use of the improvements possible in the fiber dimension of a singular set by considering a family of subsingular sets in the context required for application in sections 4 and 5. In doing this the "level preserving" property of Lemma 3.4, not required in the sequel, is lost and the form of the conclusion is drastically altered.

3.11 DEFINITION. Let $g: X \to Y$ be any embedding. If $A \subset X$ then A' denotes the set $g(A)$ and if $B' \subset Y$ then B denotes the set $g^{-1}(B)$.

3.12 LEMMA. Let M and N be compact piecewise linear manifolds of dimensions m and n, respectively, with $n - m \geq 3$. Let

$$\begin{array}{ccc} F\colon I \times (M,\partial M) \times \Delta^s & \longrightarrow & I \times (N,\partial N) \times \Delta^s \\ \downarrow & & \downarrow \\ 1\colon \Delta^s & \longrightarrow & \Delta^s \end{array}$$

be a proper piecewise linear embedding such that

$$F(\{i\} \times (M,\partial M) \times \Delta^s) \subset \{i\} \times (N,\partial N) \times \Delta^s, \quad i = 0,1.$$

Given an $\epsilon > 0$ there is a proper fibered piecewise linear ϵ ambient isotopy, φ_t, such that

(i) $\varphi_t(\{i\} \times (N,\partial N) \times \Delta^s) \subset \{i\} \times (N,\partial N) \times \Delta^s$, $i = 0,1$.

For $p = p(2,3)$, $p' = p(3)$ and $\varphi_1 \circ F$ define $S_o,\ldots,S_{t+1}$, $t = \left|\dfrac{m}{n-m-2}\right|$, by $S_o = I \times M \times \Delta^s$, $S_{2i+1} = S(p \circ \varphi_1 \circ F \mid p^{-1} \circ p(S_{2i}))$, and $S_{2i} = S(p \mid S_{2i-1})$. Then

(ii) for $1 \le i \le t$, $\dim(S_i) = \dim(S_i') \le s + m + i(m-n+2)$ and $\dim(S_{t+1}) = \dim(S_{t+1}') \le s + m - n + 2$,

(iii) for $1 \le i \le t$, $fp'\dim(S_i) = f'p\dim(S_i') \le \max(0, m + i(m-n+2))$ and $fp'\dim(S_{t+1}) = fp'\dim(S_{t+1}') \le 0$

(iv) p is nondegenerate when restricted to $\varphi_1 \circ F(I \times M \times \Delta^s)$ or S_i and S_i', $i = 1,\ldots,t+1$.

Furthermore, if (ii), (iii), and (iv) were satisfied for a subcomplex (K,K_o) of $(M,\partial M)$, and $F(M,\partial M) \cap (I \times (L,L_o) \times \Delta^s) \subset F(K,K_o)$, then φ_t may be taken to be a proper fibered piecewise linear ϵ ambient isotopy of $I \times (N,\partial N) \times \Delta^s$ relative to $F(K,K_o) \cup (I \times (L,L_o) \times \Delta^s)$.

3.13 REMARK. The process of defining the singular sets terminates with the definition of S_{t+1} as both S_t and S_{t+1} have $fp'\dim 0$. In this regard note that this process first improves the $fp'\dim$ of the singular sets until zero is reached at which point the $p'\dim$ is lowered. Dimensional improvement must then terminate.

The proof of Lemma 3.12 is simply an application of Lemma 3.4 followed by careful application of Lemma 3.8. In the proof $\overline{A}$ denotes the closure of A. Warning! A different use of this notation is employed in the remainder of the paper.

3.14 PROOF OF LEMMA 3.12. Lemma 3.4, with (K,K_o) equal to $(M,\partial N)$ and (K',K_o') equal to (K,K_o), implies the existence of a proper level preserving fibered simplicial $\epsilon/(m+1)$ ambient isotopy, φ_t^1, of $I \times (N,\partial N) \times \Delta^s$ relative to $(K,K_o) \cup (I \times (L,L_o) \times \Delta^s)$ satisfying conditions (i), (ii) and (iii).

Condition (i) of this lemma is protected by the level preserving nature of φ_t^1.

By condition (i) of Lemma 3.4, for σ and τ maximal dimensional simplices of $\varphi_1^1 \circ F(S_o)$ not in $I \times \partial N \times \Delta^s$, i.e., of dimension $1+m+s$, $\dim(p(\sigma) \cap p(\tau)) \leq 2(1+m+s) - s - n \leq s + m + (m-n+2)$. Moreover if σ and τ are in $I \times \partial N \times \Delta^s$ and therefore of dimension $1 + (m-1) + s = m+s$, $\dim(p(\sigma) \cap p(\tau)) \leq 2(m+s) - s - (n-1) \leq s + m + (m-n+1)$. This $\dim(S_1') \leq s + m + (m-n+2)$ and condition (ii) is partially verified.

Similarly, by condition (ii) of Lemma 3.4, for σ and τ maximal dimensional simplices of $\varphi_1^1 \circ F(S_o)$ not in $I \times \partial N \times \Delta^s$, $fp'\dim(p(\sigma) \cap p(\tau)) \leq \max(0,2(m+1) - n \leq \max(0,m+(m-n+2))$. If σ and τ are in $I \times \partial N \times \Delta^s$,

$\text{fp}'\dim(p(\sigma) \cap p(\tau)) \leq \max(0, 2(m) - (n-1)) \leq \max(0, m + (m-n+1))$. Thus $\text{fp}'\dim(S'_1) \leq \max(0, m + (m-n+2))$ and condition (iii) is partially verified.

Since $\varphi_1^1 \circ F$ is fiber preserving the above also implies that $\dim(S_1) \leq s + m + (m-n+2)$ and $\text{fp}'\dim(S_1) \leq \max(0, m + (m-n+2))$.

By condition (iii) of Lemma 3.4 p is nondegenerate when restricted to $\varphi_1^1 \circ F(I \times M \times \Delta^s)$, and hence S'_1, since $\text{fp}'\dim(D^1 \times M \times \Delta^s) = 1 + m \leq n$ and $\text{fp}'\dim(I \times \partial M \times \Delta^s) = m \leq n-1$.

The next stage of the argument is an application of Lemma 3.4 to $I \times (M, \partial M) \times \Delta^s$ where (K, K_o) is taken to be $((\varphi_1^1 \circ F)^{-1}(\overline{S}'_1), (\varphi_1^1 \circ F)^{-1}(\overline{S}'_1) \cap (I \times \partial M \times \Delta^s)) \cup (K, K_o)$ and (K', K'_o) taken to be (K, K_o) with L empty. By Lemma 3.4 there is a proper level preserving fibered simplicial ϵ' ambient isotopy, $\hat{\varphi}_t^2$, of $I \times (M, \partial M) \times \Delta^s$ such that conditions (i), (ii) and (iii) are satisfied. Since ϵ' can be selected as small as desired $(\varphi_1^1 \circ F)\hat{\varphi}_t^2(\varphi_1^1 \circ F)^{-1}$ may be required to be a proper fibered ϵ isotopy of $(\varphi_1^1 \circ F)(I \times M \times \Delta^s)$ in itself relative to $F(K, K_o)$. This extends to a proper fibered piecewise linear ϵ ambient isotopy φ_t^2 of $I \times N \times \Delta^s$ relative to $F(K, K_o) \cup (I \times (L, L_o) \times \Delta^s)$, cf. Hudson [39]. Furthermore $\overline{S}'_1$ is not altered (in $I \times N \times \Delta^s$), condition (i) is preserved, conditions (ii) and (iii), as they concern S'_1, are preserved, and condition (iv) as it concerns $\varphi_1^2 \circ \varphi_1^1 \circ F(I \times M \times \Delta^s)$ and S'_1, is unchanged.

Let $S_1 = (\varphi_1^2 \circ \varphi_1^1 \circ F)^{-1}(S'_1)$ and $S_2 = S(p \mid S_1)$. Conditions (ii) and (iii) will be verified for those maximal dimensional simplices of S_1 and S_2 which are not in $I \times \partial M \times \Delta^s$. This latter case is easily demonstrated by the same argument employed in the previous case. Suppose then that σ and τ are maximal dimensional simplices of S_1, i.e., of dimension $\leq s + m + (m-n+2)$. By condition (i) of Lemma 3.4 $\dim(p(\sigma) \cap p(\tau)) \leq$

$2(s+m+(m-n+2)) - s - m = s + m + 2(m-n+2)$, proving condition (ii) for S_2. Similarly $fp'\dim(S_2) \leq \max(0, 2(m+(m-n+2))-m) = \max(0, m+2(m-n+2))$, proving condition (iii) for S_2. The basic point of this computation is that if the fiber dimension of the intersections of the projected simplices is greater than zero then the dimension of its projection, via p', is equal to the smaller of the dimensions of the projections of the two simplices. This fact follows from conditions (i) and (ii) of Lemma 3.4. Thus the primary effect of the general position technique is to lower the fiber dimension until zero is reached. If this is already the case it is possible that the fiber dimension may no longer be lowered and, consequently, the dimension of the projection to Δ^s is lowered.

Condition (iii) of Lemma 3.4 implies condition (iv) of the lemma as far as S_1, and hence S_2, are concerned. The situation for S_2' would follow from the statement for S_1' except that it is necessary to make a further series of very local modifications in order to fully achieve the conditions required in (ii) and (iii). Let $\hat{S}_i = p^{-1} \circ p(S_i)$ and $\tilde{S}_i = p^{-1} \circ p(\overline{S}_i)$. One effect of the general position lemma is that each simplex of $S(p \circ \varphi_1^2 \circ \varphi_1^1 \circ F \mid \hat{S}_2) \subset \hat{S}_2 \cap S_1$ lies in the interior of a simplex of S_1. Moreover all, but certain of those with $fp'\dim$ 0, have $fcp'\dim$ at least $n-m-2 \geq 1$. It is this fact, which need not hold for $\tilde{S}_2$ in $\tilde{S}_1$, that makes the further modifications possible. During this construction the level preserving nature of the ambient isotopy is lost.

Let $I \times N \times \Delta^s$ be triangulated by a subdivision of a first derived subdivision of a triangulation in which $\tilde{S}_1'$ and $\tilde{S}_2$ are subcomplexes such that p and p' are simplicial.

List the simplices of $\tilde{S}_2$ which are in $\tilde{S}_2' \cap (\varphi_1^2 \circ \varphi_1^1 \circ F)(K)$ or in simplices of $\overline{S}_1'$ which project nondegenerately to Δ^s by $v_1, \ldots, v_k$. The remaining vertices of $\tilde{S}_2'$, $w_1,\ldots,w_\ell$, are ordered by increasing dimension of the open simplices from which they arose. Let $\tilde{S}_{2,i}'$ denote the subcomplex of $\tilde{S}_2'$ consisting of all simplices whose vertices are in the set $\{v_1,\ldots,v_k,w_1,\ldots,w_i\}$. By construction $\tilde{S}_{2,i}'$ is a full subcomplex of $I \times N \times \Delta^s$ and, therefore, of $\hat{S}_{2,j}'$, $i < j$.

As in the proof of Lemma 3.4 a proper fibered piecewise linear $\epsilon/m+1$ ambient isotopy of $I \times N \times \Delta^s$ relative to $F(K) \cup (I \times L \times \Delta^s)$ satisfying the conclusions of the lemma, as far as S_1, S_1', S_2, S_2', and S_3' are concerned, is constructed by induction with respect to the $\hat{S}_{2,i}'$.

Assume that such an $(i-1)\epsilon/\ell(m+1)$ ambient isotopy has been constructed for $\hat{S}_{2,i-1}'$. The inductive process begins easily with the case $i = 0$ since the conclusion is satisfied by hypothesis for those vertices in $\tilde{S}_2' \cap (\varphi_1^2 \circ \varphi_1^1 \circ F)(K)$. For those of $\overline{S}_1'$ which lie in simplices projecting nondegenerately to Δ^s the conclusion is satisfied since all simplices of $\tilde{S}_2'$ having some of these as vertices lie in such a simplex of $\overline{S}_1'$ ($\tilde{S}_2'$ is a full subcomplex). Therefore the general position conditions of the lemma follow from the fact that the dimensions of such simplices are no larger than $s + m - n + 2$ and their fiber dimensions are precisely zero.

By hypothesis the vertex, $p(w_i)$ lies in the interior of a simplex $p(\overline{S}_1')$, σ, in some prior triangulation for which p was simplicial and $(\varphi_1^2 \circ F)(I \times M \times \Delta^s)$ and $\overline{S}_1'$ were full. Consider the "affine subspaces" of σ, σ having the usual linear structure of a simplex, which are determined by the simplices of $p(\tilde{S}_{2,i-1}' \cap p^{-1}(\sigma))$. Since all such simplices have fiber codimension at least one in σ and since all simplices of $\hat{S}_{2,i}'$ not in $\hat{S}_{2,i-1}$ lie, with fiber codimension at least one, in that simplex of $p^{-1}(\sigma,$ γ, containing w_i Lemma 3.8 implies that there is a point w_i' arbitrarily close to w_i in $(p')^{-1}(w_i) \cap \gamma$ such that the linear translation of w_i to

w_i' in γ, relative to $\partial\gamma$, followed by p gives the required fibered general position properties for the simplices of $\hat{S}_{2,i}'$.

For example suppose that α and β are maximal dimensional simplices of $\hat{S}_{2,i}'$, i.e., of dimension $s+m+2(m-n+2)$ and fiber dimension $m+2(m-n+2)$, whose projections lie in a $s+m+(m-n+2)$ simplex of S_1' with fiber codimension $m-n+2$. Then $\dim(p(\alpha)\cap p(\beta)) \le s+m+3(m-n+2)$, $fp'\dim(p(\alpha)\cap p(\beta)) \le \max(0, m+3(m-n+2))$. The same argument proves the statement that such intersections have fiber codimension at least one in $\tilde{S}_2'$ for simplices of less than maximal possible dimension as indicated above. This fact then provides the inductive key to the construction of singular sets having the required properties.

The desired $i\epsilon/\ell(m+1)$ isotopy is achieved by taking the composition of that isotopy defined for the case $i-1$ with an $\epsilon/\ell(m+1)$ ambient isotopy of $I\times N\times\Delta^s$, whose support is the star of w_i in $I\times N\times\Delta^s$, by taking w_i' to w_i and by extending the "translation" in γ to the star. This isotopy leaves $F(K)\cup(I\times L\times\Delta^s)$ fixed and leaves the simplices of $(\varphi_1^2\circ\varphi_1^1\circ F)(I\times M\times\Delta^s)$ and S_1' in that previous triangulations setwise invariant since S_1' and $\hat{S}_2'$ were full subcomplexes of that triangulation.

Thus, by induction, there is a proper fibered piecewise linear $\epsilon/m+1$ ambient isotopy of $I\times(N,\partial N)\times\Delta^s$ relative to $F(K,K_o)\cup(I\times(L,L_o)\times\Delta^s)$, φ_t^3, leaving the simplices of $(\varphi_1^2\circ\varphi_1^1\circ F)(I\times M\times\Delta^s)$ and $\overline{S}_1'$ setwise invariant in the triangulation employed in the definition of S_1', such that $(\varphi_1^3\circ\varphi_1^2\circ\varphi_1^1\circ F)$ satisfies the conclusions of the lemma as far as S_1, S_1', S_2, S_2', and S_3' are concerned.

As things stand S_4 need not satisfy the conclusions so that an isotopy of $I\times M\times\Delta^s$ relative to K is defined which leaves S_1 and the simplices of $\tilde{S}_2$ setwise invariant and as in the definition of φ_t^3, puts the projections of the simplices of S_3 in general position. Then, just as in the definition of φ_t^2, this isotopy is realized as an ambient isotopy of $I \times N \times \Delta^s$, etc. Since the simplices of S_3' have fiber codimension at least one in S_2', if their fiber dimension is larger than zero, the dimensional properties of $S_4 = S(p \mid S'_3 = (\varphi_1^4 \circ \varphi_1^3 \circ \varphi_1^2 \circ \varphi_1^1 \circ F)^{-1}(S_3))$ are easily verified as above. Moreover the simplices of S_4 having fiber dimension greater than zero have fiber codimension at least $n - m - 2$ in S_3.

Thus inductively a family of $m+1$ $\epsilon/m+1$ ambient isotopies are constructed so that $\varphi_t = \varphi_t^{m+1} \circ \cdots \circ \varphi_t^1$ satisfies the conclusions of the lemma.

3.15 REMARK. The method of proof indicates that the usual general position theorems, cf. Zeeman [73] or Hudson [38], can be extended to fibered general position theorems for complexes as piecewise linear spaces. Unfortunately, as indicated by Lemma 3.4, the appropriate conclusions of such generalizations take on a rather complicated form.

SECTION 4. FIBERED SUNNY COLLAPSING

In Section 5 a variant of Zeeman's sunny collapsing will be required in the proof of the local modification lemma. To this end two fibered sunny collapsing procedures are considered in this section. The first is more closely related to Zeeman's but seems unsuitable in the present situation. The second does not provide a terminal collapsing procedure but is adequate to accomplish the requirements in the next section. Before giving these lemmas it is necessary to extend the classical definitions to the present situation.

4.1 DEFINITION. If $X \subset I \times Y$ the shadow of X in $I \times Y$ is the set $sh(X) = \{(t,y) \in I \times Y \mid \exists\, s \in I \ni s > t$ and $(s,y) \in X\}$.

4.2 DEFINITION. If $K' \subset K \subset I \times L \times \Delta^s$ are simplicial complexes and subcomplexes such that

(i) $p(2,3)$ and $p(3)$ are simplicial on $I \times L \times \Delta^s$,

(ii) there is a simplex α, the join of a vertex v and a simplex β in K, such that $p(3)(v)$ is a vertex of $p(3)(\beta)$,

(iii) $\text{closure}_K(K \setminus K') \cap K' = v(\partial\beta)$

((iv) $sh(K) \cap \alpha \subset v(\partial\beta) \cup \alpha)$,

we say that $K \searrow^{es} K'$ is an elementary fibered simplicial (sunny) collapse.

4.3 DEFINITION. If $K' \subset K \subset I \times L \times \Delta^s$ are simplicial complexes and subcomplexes such that

(i) $p(2,3)$ and $p(3)$ are simplicial on $I \times L \times \Delta^s$,

(ii) there is a subcomplex $I \times \alpha$, α a simplex, triangulated such that $p\colon I \times \alpha \to \alpha$ is simplicial,

(iii) $\text{closure}_K(K \setminus K') \cap K' = I \times \partial\alpha \cup \{0\} \times \alpha$

((iv) $sh(K) \cap I \times \alpha \subset I \times \partial\alpha \cup \{0\} \times \alpha$),

we say that $K \searrow^{ec} K'$ is an elementary fibered cylindrical (sunny) collapse.

4.4 DEFINITION. A fibered simplicial {cylindrical} (sunny) collapse of K to K_o in $I \times L \times \Delta^s$ is a sequence of elementary fibered simplicial {cylindrical} (sunny) collapses

$$K = K_r \searrow K_{r-1} \searrow \cdots \searrow K_1 \searrow K_o .$$

Evidently an elementary fibered cylindrical (sunny) collapse may be achieved by a fibered simplicial (sunny) collapse. Thus this is also true of a fibered cylindrical (sunny) collapse. The cylindrical collapsing is employed in the second lemma while the simplicial collapsing is used in the generalization of Hudson's sunny collapsing lemma.

4.5 DEFINITION. A subcomplex $(J,J_o) \subset I \times (K,K_o) \times \Delta^s$ is said to be vertical modulo J_o if $(t,x) \in \mathrm{closure}_J(J \setminus J_o)$ implies that $I \times \{x\} \subset \mathrm{closure}_J(J \setminus J_o)$.

4.6 LEMMA. Let (K,K_o) and (L,L_o) be simplicial complexes, (J,J_o) a vertical subcomplex, modulo J_o, of $I \times (K,K_o) \times \Delta^s$, and

$$\begin{array}{ccc} F\colon (J,J_o) & \longrightarrow & I \times (L,L_o) \times \Delta^s \\ {\scriptstyle p(3)\,|\,(J,J_o)}\big\downarrow & & \big\downarrow{\scriptstyle p(3)} \\ 1\colon \Delta^s & \longrightarrow & \Delta^s \end{array}$$

a proper simplicial fibered embedding such that

(i) $S(p(2,3) \circ F)$ has local fiber codimension at least two in J modulo J_o, and

(ii) $p(2,3) \mid S(p(2,3) \circ F)$ is nondegenerate modulo J_o.

Then there is a fibered simplicial sunny collapse of $F(J)$ to $F(J_o \cup (J \cap \{0\} \times K \times \Delta^s))$ in $I \times L \times \Delta^s$, for some subdivision of $F(J)$.

4.7 COROLLARY. Let (K,K_o) and (L,L_o) be simplicial complexes and

$$\begin{array}{ccc} F\colon I \times (K,K_o) \times \Delta^s & \longrightarrow & I \times (L,L_o) \times \Delta^s \\ \Big\downarrow p(3) & & \Big\downarrow p(3) \\ 1\colon \Delta^s & \longrightarrow & \Delta^s \end{array}$$

be a proper simplicial fibered embedding such that

(i) $S(p(2,3) \circ F)$ has local fiber codimension ≤ 2 in $I \times K \times \Delta^s$, modulo $I \times K_o \times \Delta^s$, and

(ii) $p(2,3) \mid S(p(2,3) \circ F)$ is nondegenerate modulo $I \times K_o \times \Delta^s$. Then there is a fibered simplicial sunny collapse of $F(I \times K \times \Delta^s)$ to $F((I \times K_o \times \Delta^s) \cup \{0\} \times K \times \Delta^s)$ in $I \times L \times \Delta^s$, for some subdivision of $F(I \times K \times \Delta^s)$.

4.8 REMARK. Corollary 4.7 follows from Lemma 4.6 by taking (J,J_o) to be equal to $I \times (K,K_o) \times \Delta^s$. The necessity of stating Lemma 4.6 in its present form stems from the inductive nature of its proof. This would not be possible in the form given in the corollary.

4.9 REMARK. Let $(N,\partial N)$ be a compact piecewise linear manifold and let $F\colon I \times (K,K_o) \times \Delta^s \to I \times (N,\partial N) \times \Delta^s$ be a proper fibered piecewise linear embedding such that $\dim(N,\partial N) - \dim(K,K_o) \geq 3$. Conditions (i) and (ii) in the hypothesis of Corollary 4.7 can be achieved by an application of Lemma 3.5, i.e., there is an arbitrarily small proper level preserving fibered piecewise linear ambient isotopy φ_t of $I \times (N, \partial N) \times \Delta^s$ such that $\varphi_1 \circ F$ satisfies conditions (i) and (ii). Condition (i) is a consequence of condition (ii) of Lemma 3.5 while condition (ii) follows directy from condition (iii).

The proof of Lemma 4.6 is an extension of the methods employed by Hudson to prove an analogous lemma, cf. Lemma 5.4 of [39] or Lemma 9.5 of [38]. As with most collapsing arguments the method of proof indicates that there are manifold versions and extensions of the lemma given here. This collapsing lemma is included primarily as an indication of the manner in which various techniques can be extended to the fibered situation.

4.10 DEFINITION. In a simplicial complex $I \times L$, with $p(2)\colon I \times L \to L$ the simplicial projection, a simplex α is below a simplex β if $\alpha \cap \text{closure}(\text{sh}(\beta, D^1 \times L)) \not\subset \partial\alpha$.

4.11 PROOF OF THEOREM 4.6. The proof is by a double induction, first with respect to s and then with respect to $\dim(J) = k$. The induction can begin with the case $s = 0$ and arbitrary k. This case is a straightforward extension of the standard sunny collapsing lemma (the subcomplex J_o is merely ignored in the collapsing process) cf. Zeeman [73] and Hudson [38].

The induction for a fixed s can begin with $k = 0$ so that $S(p(2,3) \circ F) = S$ is empty. This case is easily accomplished since (J, J_o) consists of a disjoint collection of complexes of the form $I \times \{x\} \times (\Delta, \Delta_o)$ where Δ is a subcomplex of Δ^s. It is sufficient to consider each such subcomplex $I \times \{x\} \times (\Delta, \Delta_o)$ and $\{x\} \times (\Delta, \Delta_o)$ so that $p(2,3)$ is simplicial. Let $\alpha_1, \ldots, \alpha_\ell$ denote the s simplices of $\{x\} \times \Delta$ not in $\{x\} \times \Delta_o$ and $\beta_{1,t}, \ldots, \beta_{m_t,t}$ the $s+1$ simplices of $I \times \{x\} \times \Delta$ which project onto α_t

ordered so that if $i < j$ then $\beta_{j,t}$ is not below $\beta_{i,t}$.

This inductive stage will be verified by showing that for each $I \times \{x\} \times \alpha_t$ there is a fibered simplicial sunny collapse to $(I \times \{x\} \times (\partial\alpha_t)) \cup (\{0\} \times \{x\} \times \alpha_t)$. Indeed all that is required is a fibered simplicial collapse since $S = \emptyset$. This is accomplished in order of decreasing i for fixed t. There is an elementary fibered simplicial collapse of $(I \times \{x\} \times (\partial\alpha_t)) \cup (\cup_{j=1}^{i} \beta_{j,t})$ to $(I \times \{x\} \times (\partial\alpha_t)) \cup (\cup_{j=1}^{i} \beta_{j,t})$, or to $(I \times \{x\} \times (\partial\alpha_t)) \cup (\{0\} \times \{x\} \times \alpha_t)$ if $i = j$, since $\beta_{i,t} = v\gamma$ with $p(\gamma) = \alpha_t$ and v a vertex below some vertex of γ. Since $\beta_{j,t}$ is below γ if and only if $j \leq i$ and $v(\partial\gamma)$ is contained in the next collapsed complex this is an elementary fibered simplicial collapse, cf. figure 4.12. When these collapses have been accomplished, for all $\alpha_1, \ldots, \alpha_\ell$, this inductive stage is completed by employing the inductive hypothesis, i.e., the case for $s-1$.

Assume then that the lemma has been verified for all dimensions of K and Δ^t, $t < s$, as well as for all dimensions less than k and Δ^s. Triangulate (J, J_o), $(J, J_o) \cap (\{0\} \times (K, K_o) \times \Delta^s)$, and Δ^s so that $p(2,3)$ and $p(3)$ are simplicial and $S = S(p(2,3) \circ F)$ is a subcomplex. Subdivide S so that $p(2,3) \circ F|S$ is simplicial and note that $p(3) | S$ is still simplicial with respect to this new triangulation of S since $p(3) = p(3) \circ F$. Also note that $S \cap (\{0\} \times K \times \Delta^s)$ has been subdivided.

It may be assumed that Δ^s has not been subdivided for once this case has been proved, relative $(J, J_o) \cap (I \times (K, K_o) \times \partial\Delta^s)$ the collapsing relative the $s-1$ skeleton of a subdivided Δ^s follows and then, by the inductive hypothesis, the lemma is proved. The case for a single simplex Δ^s is proved, as indicated above, by induction on $k = \dim(K)$.

Let $\alpha_1, \ldots, \alpha_\ell$ denote the $k+s$ simplices of $J \cap (\{0\} \times K \times \Delta^s)$ which are not in $J_o \cap (\{0\} \times K \times D^s)$. Since S has local fiber codimension at least two, modulo J_o, the maximum dimension of simplices of S not in J_o is $k-1+s$ and the maximum fiber dimension is $k-1$. Let $\beta_1, \ldots, \beta_m$ be these simplices of S in the subdivision such that $p(2,3) \circ F$ is simplicial, which are not in $J_o \cup (J \cap (\{0\} \times K \times \Delta^s))$, and ordered so that if $i < j$ then β_j is not below β_i.

Since, prior to subdivision, S was a subcomplex of J each β_i lies in a $k-1+s$ face, having fiber dimension $k-1$, of a $k+1+s$ simplex. As $p(2,3)$ is simplicial and $p(3) \mid \beta_i$ is simplicial (to Δ^s) each β_i is contained in $I \times \alpha_j^i$, where α_j^i is a $k-1+s$ face of α_j having fiber dimension $k-1$ and (hence) projective dimension s. Furthermore $p(2,3) \mid s$ is nondegenerate, modulo J_o, so that the projection of β_i to α_j^i is nonsingular but not necessarily simplicial, cf. figure 4.13 where S is the union of β_1, β_2, and the shaded 2 simplex which lies in J_o.

"Fibered blisters" analogous to those described by Hudson will next be constructed and finally employed in the description of the fibered simplicial sunny collapsing. Let b_i be the barycenter of β_i. Since $p(2,3)$ is nonsingular on β_i there is an x_i directly below and arbitrarily close to b_i (as $b_i \notin \{0\} \times K \times \Delta^s$). Similarly, if $b_i \notin \{1\} \times K \times \Delta^s$, there is a y_i directly above and arbitrarily close to b_i but not in $\{1\} \times K \times \Delta^s$. Furthermore, since $p(2,3)(\beta_i) \subset \alpha_j^i$, there is a $z_i \in I \times (\text{int } \alpha_j)$ arbitrarily close to b_i and such that $p(1,3)(z_i) = p(1,3)(b_i)$.

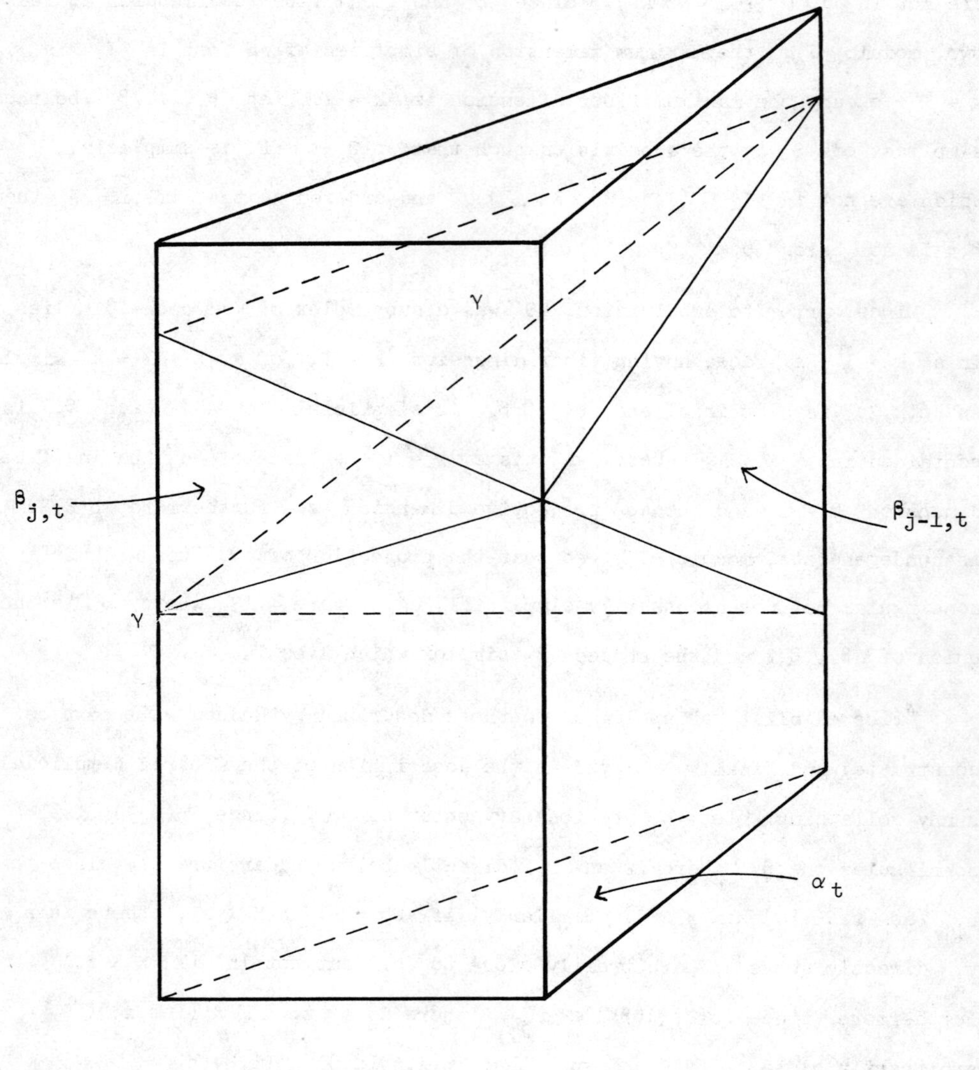

4.12 COLLAPSING WHEN $k = 0$.

Let

$$\beta_i = \begin{cases} z_i x_i y_i(\partial \beta_i) & p(1)(b_i) \neq 1 \\ \\ z_i x_i \beta_i & p(1)(b_i) = 1 \end{cases}$$

The x_i, y_i and z_i can be selected so that $B_i \cap B_j = \beta_i \cap \beta_j$, $B_i \cap (\{0\} \times K \times \Delta^s) = \beta_i \cap (\{0\} \times K \times \Delta)$, $B_i \cap (I \times K_o \times \Delta^s) = \beta_i \cap (I \times K_o \times \Delta^s)$, and if $p(1)(b_i) < 1$ then $B_i \cap (\{1\} \times K \times \Delta^s) = \beta_i \cap (\{1\} \times K \times \Delta^s)$, cf. figure 4.14 .

The fibered simplicial sunny collapse is accomplished in the same manner as employed by Hudson. First each of the $I \times \alpha_j$'s is collapsed to the subcomplex $(I \times \partial\alpha_j) \cup (\{0\} \times \alpha_j) \cup (\bigcup_{\alpha_j^i} B_i)$. All that is required is a fibered simplicial collapse since the singular set does not intersect that which is to be collapsed. This, in turn, follows from the observation that there is a particularly nice fibered piecewise linear homeomorphism of the closure of the complement of $(I \times \partial\alpha_j) \cup (\{0\} \times \alpha_j) \cup (\bigcup_{\alpha_j^i} B_i)$ in $I \times \alpha_j$ to $I \times \alpha_j$ and a straightforward extension of the simplicial collapsing procedure described in the proof of the inductive step for arbitrary s and $k = 0$ to the case $I \times \alpha_j \overset{s}{\searrow} I \times \partial\alpha_j \cup \{0\} \times \alpha_j$. The desired homeomorphism can be described as follows: When selecting z_i choose also a z_i' collinear with b_i and z_i and slightly beyond z_i. The composition of the extensions of the fibered piecewise linear homeomorphisms taking $z_i z_i' x_i y_i(\partial\beta_i)$ or $z_i z_i' x_i \beta_i$ to $b_i z_i' x_i y_i(\partial\beta_i)$ or $b_i z_i' x_i \beta_i$, respectively, by the identity on the remainder of the complement is the desired homeomorphism, cf. figure 4.10.

The second stage of the process collapses each of the B_i in order of decreasing index so as to insure that it is a sunny collapse. Specifically, let

$$C_i = \begin{cases} z_i y_i(\partial\beta_i) \cup z_i x_i(\partial\beta_i) & p(1)(b_i) < 1 \\ \\ z_i x_i(\partial\beta_i) & p(1)(b_i) = 1 \end{cases}$$

There is a fibered simplicial collapse of each B_i to the corresponding C_i given as follows: if $p(1)(b_i) < 1$,

$$B_i = z_i x_i y_i(\partial\beta_i) \searrow^{es} z_i y_i(\partial\beta_i) \cup z_i y_i \beta_i \searrow^{es}$$

$$z_i y_i(\partial\beta_i) \cup z_i x_i(\partial\beta_i) \cup z_i \,_i \searrow^{es} z_i y_i(\partial\beta_i) \cup z_i x_i(\partial\beta_i),$$

and if $p(1)(b_i) = 1$,

$$B_i = z_i x_i \beta_i \searrow^{es} z_i x_i(\partial\beta_i) \cup z_i \beta_i \searrow^{es} z_i x_i(\partial\beta_i).$$

Thus collapsing each of the B_i, in order of decreasing index, defines a fibered simplicial sunny collapse.

The net result of the sequence of collapses described above is a fibered simplicial sunny collapse of J modulo J_o to

$$[\tilde{p}^{-1}(\tilde{p})(J^{k-2})) \cup (J \cap (\{0\} \times K \times \Delta^s)) \cup J_o \cup (\bigcup_{i=1}^{m} C_i)] \setminus (\bigcup_{i=1}^{m} \operatorname{int}(x_i y_i(\partial\beta_i))),$$

where $\tilde{p} = p(2,3)$.

There is a fibered piecewise linear homeomorphism from this latter complex to $\tilde{p}^{-1}(\tilde{p})(J^{k-2})) \cup (J \cap (\{0\} \times K \times \Delta^s)) \cup J_o$ whose support is $\bigcup_{i=1}^{m} C_i$. This homeomorphism is precisely that defined on the C_i as follows: $C_i = z_i y_i(\partial\beta_i) \cup z_i x_i(\partial\beta_i)$ goes to $b_i y_i(\partial\beta_i) \cup b_i x_i(\partial\beta_i)$, and if $p(1)(b_i) = 1$, then $C_i = z_i x_i(\partial\beta_i)$ goes to $b_i x_i(\partial\beta_i)$.

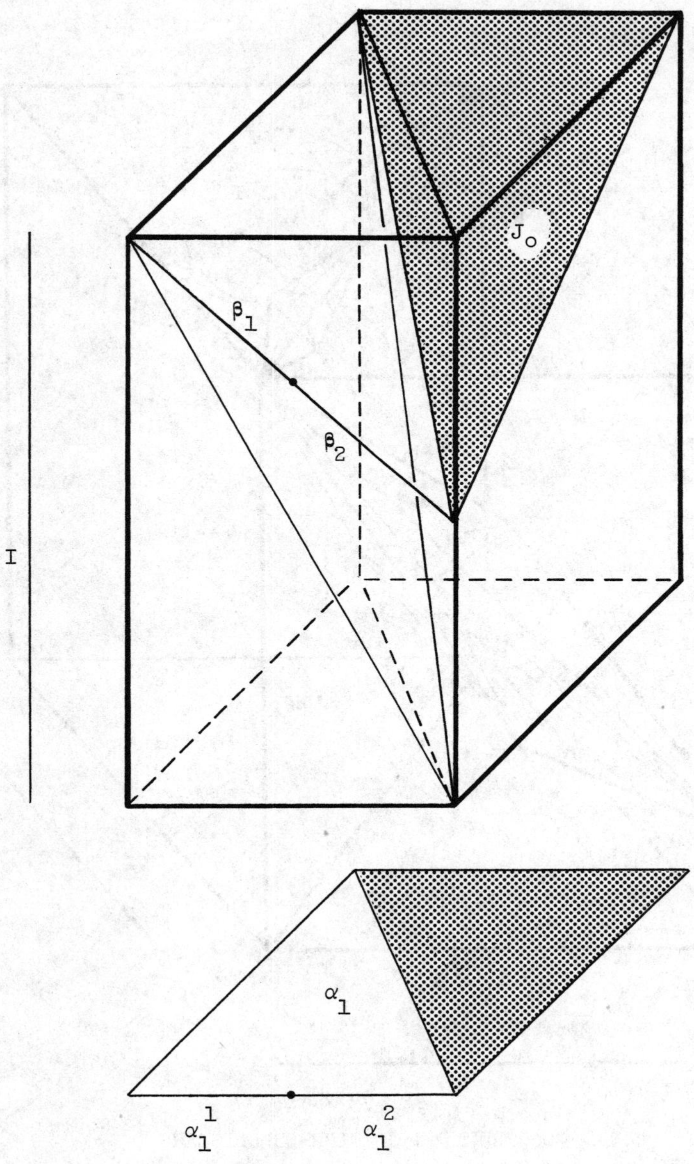

4.13 (J, J_o), AN EXAMPLE.

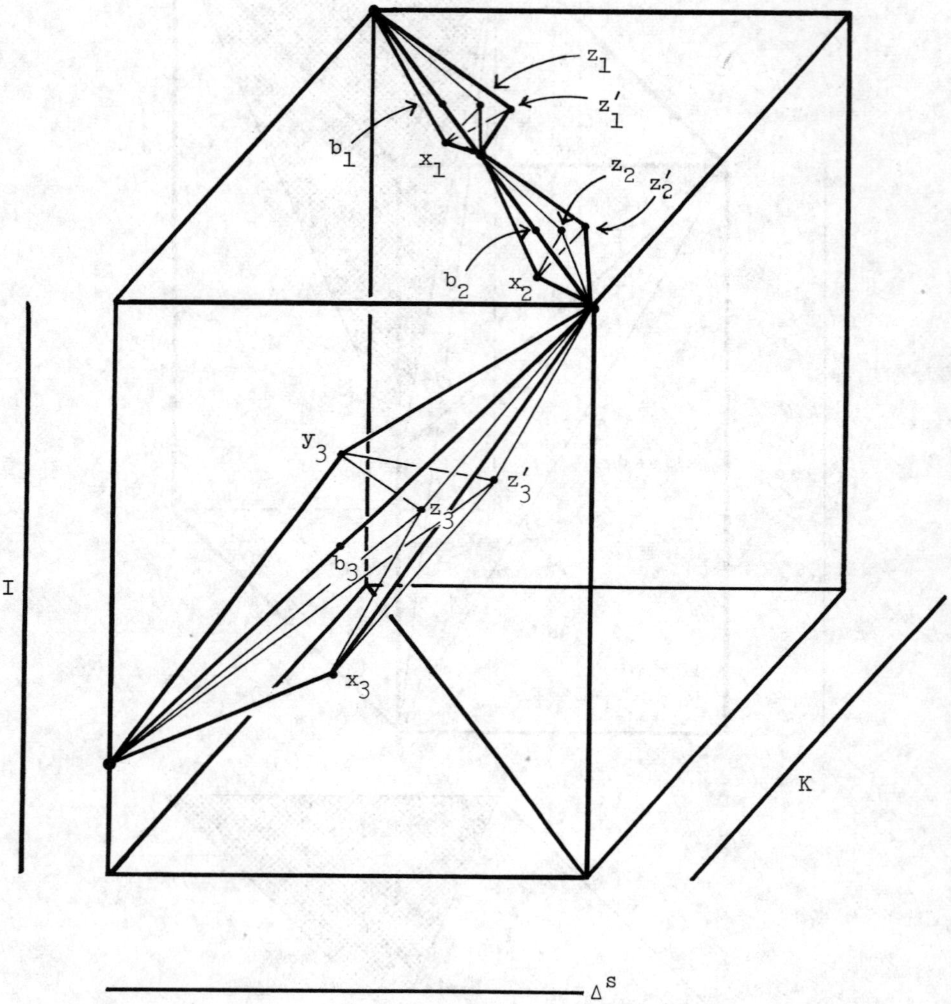

4.14 CONSTRUCTION OF FIBERED BLISTERS

The proof of the lemma is easily concluded by observing the subcomplex

$$[\tilde{p}^{-1}(\tilde{p}\,(J^{i-1})) \cup J_o \cup (\bigcup_{i=1}^{m} C_i)] \setminus (\bigcup_{i=1}^{n} \operatorname{int}(x_i y_i(\partial \beta_i))) = J$$

is fibered piecewise linearly homeomorphic, via G, to a vertical subcomplex of $I \times (K^{k-1} \cup K_o, K_o) \times \Delta^s$, that the dimension of K modulo K_o is $k-1$, and that the inductive hypotheses are satisfied for $(G(J'), J_o)$ and $F \circ G^{-1}$. First $S(\tilde{p} \circ F \circ G^{-1})$ has local fiber codimension at least two since it is contained in $S(\tilde{p} \circ F) \mid (\bigcup_{i=1}^{m} \operatorname{int} \beta_i)$, which has local fiber codimension at least two in $\tilde{p}^{-1}(\tilde{p}(J^{k-1}))$. Finally $\tilde{p} \circ F \circ G^{-1}$ is the restriction of a nondegenerate map and is therefore nondegenerate. This completes the proof of Lemma 4.6.

4.15 REMARK. Although Corollary 4.7 is the appropriate extension of Hudson's technique it unfortunately does not seem to provide the necessary control required for the engulfing lemma of Section 5. The following lemma gives, as a corollary, a collapsing procedure with the required control but does not provide a terminal collapsing. This fact stems from the limitations of the fibered general position lemma, Lemma 3.13, in the fiber dimension zero case. Thus the rather complicated statement follows from this and the desire to provide an inductive proof, as with Lemma 4.6. The residual effect of this nonterminal collapsing is the introduction of a dimensional restriction in the hypothesis of the engulfing lemma.

4.16 LEMMA. Let (K, K_o) and (L, L_o) be simplicial complexes, (J, J_o) a proper vertical subcomplex, modulo J_o, of $I \times (K, K_o) \times \Delta^s$, $p = p(2,3)$, $p' = p(3)$, and

$$\begin{array}{ccc} F\colon (J,J_o) & \longrightarrow & I \times (L,L_o) \times \Delta^s \\ {\scriptstyle p'|(J,J_o)}\downarrow & & \downarrow{\scriptstyle p'} \\ 1\colon \Delta^s & \longrightarrow & \Delta^s \end{array}$$

be a proper simplicial fibered embedding.

Let $S_o' = I \times L \times \Delta^s$ so that $S_o = J$, cf. Definition 3.11. Let $S_{2j+1} = S(p \circ F \mid p^{-1} \circ p(S_{2j}))$ and $S_{2j} = S(p \mid S_{2j-1})$ be inductively defined subcomplexes S_i, $0 \le i \le t$. Furthermore let

$$h_j = \begin{cases} p^{-1} \circ p, & j \text{ odd} \\ F^{-1} \circ p^{-1} \circ p \circ F, & j \text{ even} \end{cases}$$

and define associated subcomplexes $\mathcal{S}_i = (\prod_{j=1}^{i} h_j)(S_i)$, $0 \le i \le t$.

There is a fibered cylindrical sunny collapse of $F(J)$ to

$$F(J_o \cup (J \cap \{0\} \times K \times \Delta^s) \cup \mathcal{S}_t) \quad \text{in} \quad I \times L \times \Delta^s .$$

4.17 REMARK. If M and N are compact piecewise linear manifolds of dimensions m and n, respectively, with $n - m \ge 3$, and

$$F\colon I \times (M, \partial M) \times \Delta^s \longrightarrow I \times (N, \partial N) \times \Delta^s$$

is a proper fibered piecewise linear embedding with singularity sets, $\{S_i\}_{0 \le i \le t+1}$, particularly nice dimensional properties can be achieved by the application of Lemma 3.12. This provides the first step in the proof of the main result.

4.18 PROOF OF LEMMA 4.16. The proof is by a double induction, first with respect to s and then with respect to t. Since the case $t = 0$ is vacuous for all s assume the theorem is proved for all s and $t' < t$ and for t and all $s' < s$. Let there be given a fibered cylindrical sunny collapse of $F(J)$ to $F(J_o \cup (J_o \cap \{0\} \times K \times \Delta^s) \cup \mathcal{S}_{t-1})$ in a triangulation such that the singularity sets $\{S_i\}$ and the associated sets $\{\mathcal{S}_i\}$ are subcomplexes. Subdivide so that p is simplicial and note that $p\colon (\mathcal{S}_{t-1}, \mathcal{S}_t) \to (p(\mathcal{S}_{t-1}), p(\mathcal{S}_t))$ is a proper map.

Since all the S_i are subcomplexes and $p \mid S_i$ is nondegenerate, one can assign to each simplex α in a subdivision of $p(\mathcal{S}_{t-1} \setminus (\mathcal{S}_t \cup J_o))$ a degree as follows: $\deg(\alpha) = \max\{j \mid (I \times \operatorname{int} \alpha) \cap S_j \neq \emptyset\}$. Note that this is also the largest j such that $\alpha \subset p(S_j)$.

Let $\mathcal{S}^i_{t-1} = \mathcal{S}_t \cup \bigcup\{p^{-1}(\alpha) \mid \deg(\alpha) \geq i\} = \mathcal{S}_t \cup (\mathcal{S}_{t-1} \cap p^{-1} \circ p(S_i))$ and note that $p(\mathcal{S}^i_{t-1}, \mathcal{S}^j_{t-1}) \to (p(\mathcal{S}^i_{t-1}), p(\mathcal{S}^j_{t-1}))$, $i < j$, is a proper map. From the construction it is easily seen that $\mathcal{S}^t_{t-1} = \mathcal{S}_t$. A series of triangulations of the $\mathcal{S}^i_{t-1}$ will be given so as to facilitate the description of the collapsing process. A word of warning is appropriate since the collections of simplices will not be subcomplexes. Nonetheless they "fit together" sufficiently well for the requirements of the description.

To begin subdivide further so that in addition to the above $p \circ F$ is a simplicial map. Let $\{{}_i\alpha^{t-1}_{t-1}\}$ denote the simplices of $p(\mathcal{S}^{t-1}_{t-1}) \setminus p(\mathcal{S}^t_{t-1})$ in a subdivision of the previous triangulation such that p is simplicial. Note that the singularity sets were subcomplexes in previous triangulations and therefore remain so in this and succeeding subdivisions.

Subdivide further so that $p \circ F$ is again simplicial and let $\{{}_i\beta^{t-1}_{t-1}\}$ denote the simplices of $p \circ F(\mathcal{S}^{t-1}_{t-1} \setminus \mathcal{S}^t_{t-1})$. Next, let $\{{}_i\alpha^{t-2}_{t-1}\}$ denote the simplices of $p(S^{t-2}_{t-1}) \setminus p(S^{t-1}_{t-1})$ in a subdivision of this triangulation such that p is simplicial.

Continuing in this manner collections of simplices $\{{}_i\alpha^j_{t-1}\}$, $0 \le j \le t-1$, are defined by requiring that they be simplices of $p(S^j_{t-1}) \setminus p(S^{j+1}_{t-1})$ in the subdivision of S^j_{t-1} such that p is simplicial. Similarly collections of simplices $\{{}_i\beta^j_{t-1}\}$, $0 \le j \le t-1$, of $p \circ F(S^j_{t-1} \setminus S^{j+1}_{t-1})$, in a subdivision such that $p \circ F$ is simplicial, are constructed. These are related by the requirement, as indicated above, that each subdivision is a subdivision of the preceding triangulation. The triangulation from which this process begins shall be called the original triangulation. All succeeding subdivisions will be referred to via the collections, $\{\alpha$'s$\}$ or $\{\beta$'s$\}$, which it determines.

Since $p \circ F | S_1$ is nondegenerate there are over each ${}_i\beta^j_{t-1}$ a finite collection of simplices of the image under F of the original singular set, i.e., whose projections contain ${}_i\beta^j_{t-1}$ in their interior, which then can be numbered $1,2,3,\ldots,\ell$ in order beginning at the top in $I \times L \times \Delta^s$ and proceeding downward.

4.19 LEMMA. $F(I \times \alpha) \cap (I \times \beta)$ meets the interior of the image under F of at most one simplex in the original singularity set.

PROOF. Suppose that it meets two or more. First their images under $p \circ F$ must be equal as the image of their interiors meet and $p \circ F$ was simplicial. Second α is contained in the interior of their projections, under p. Thus $\deg(\alpha)$ is not even as α is in the interior of a simplex, γ, such that $p \circ F | (I \times \gamma)$ is singular. This also shows that $\deg(\alpha)$ cannot be odd as the projections of these simplices contain α.

4.20 REMARK. The key observations are that $p \circ F \mid (S_{2j} \setminus S_{2j+1})$ is nonsingular and that, recalling

$$S_{2j+1} = S(p \circ F \mid S_{2j}),$$
$$p\colon S_{2j+1} \setminus S_{2j+2} \longrightarrow I \times L \times \Delta^s$$

is nonsingular.

It may be assumed that the collections $\{{}_i\beta^j_{t-1}\}$ and $\{{}_i\alpha^j_{t-1}\}$, $0 \le j \le t-1$, are indexed uniquely by the i's, i.e., there do not exist two β's or two α's with the same first index. Let $\ell_i(\alpha)$, the level of α over the ith β, be the number of simplex, in image under of F of the original singularity set, which contains $F(I \times \alpha)$ and whose projection contains ${}_i\beta^j_{t-1}$ in its interior. If there does not exist such a simplex the level is defined to be zero. By the previous lemma $\ell_i(\alpha)$ is well defined. Let the level of α be defined as the tuple of levels of α with respect to the ${}_i\beta$'s, i.e., $\ell(\alpha) = (\ell_1(\alpha), \ell_2(\alpha), \dots, \ell_i(\alpha), \dots)$.

4.21 LEMMA. All the nonzero levels of α, for which the corresponding levels of α' are also nonzero, are equal. Moreover, the levels arise from a single simplex in the original singularity set.

PROOF. Without loss, assume that $\deg(\alpha) \le \deg(\alpha')$ and that $\ell_s(\alpha) \ne 0 \ne \ell_s(\alpha')$ and $\ell_t(\alpha) \ne \ell_t(\alpha')$. If the simplices in $I \times \alpha$ corresponding to $\ell_s(\alpha)$ and $\ell_t(\alpha)$ are contained within a single simplex of the original singularity set their levels are equal. Thus assume that they like in distinct simplices (for $\ell_s(\alpha)$ and $\ell_t(\alpha)$). Their projections must contain α since $\ell_s(\alpha) \ne \ell_t(\alpha)$. From the previous remark $\deg(\alpha)$ cannot be even as $I \times \alpha$ contains singularities in that degree, nor can $\deg(\alpha)$ be odd as both singularities project onto α. Thus $\ell_s(\alpha) = \ell_t(\alpha)$ with levels arising from a single simplex in the original singularity set.

To complete the proof of the lemma it is sufficient to prove that, in this case, $\ell_s(\alpha') = \ell_t(\alpha')$ as well. As with α, assume that the corresponding simplices lie within distinct simplices in the original singularity set. Their projections must contain α' since the levels are unequal. Both images under F have the same projection as that of the simplex giving the levels for α. Furthermore, since $\deg(\alpha) \le \deg(\alpha')$, the images, under $p \circ F$, of their restrictions have nonempty intersection as it must contain the image of the singularity arising from $I \times \alpha$. That this is the case follows from the subdivision construction of the simplices $\{{}_i\alpha^j_{t-1}\}$.

Thus $\deg(\alpha')$ cannot be even as $p \circ F | I \times \alpha$ has singularities. $\mathrm{Deg}(\alpha')$ cannot be odd since these singularities arise from simplices in the original singular set whose projections contain α'. Therefore $\ell_s(\alpha') = \ell_t(\alpha')$ and the levels arise from a single simplex in the original singularity set. This proves the lemma.

The next step in the argument is the construction of a partial order (leading to a related total order required for the sunny collapsing) by employing the levels associated to the $\{{}_i\alpha^j_{t-1}\}$. Thus let $\alpha \circledcirc\!\!\le \alpha'$ if and only if $\ell(\alpha)$ and $\ell(\alpha')$ have nonzero coordinates in common and those of α are no larger than those of α'. Let $\le$ denote the smallest preorder of $\{{}_i\alpha^j_{t-1}\}$ extending this relation. To show that $\le$ is a partial order it suffices to show that $\le$ is antisymmetric, i.e., $\alpha \le \alpha' \le \alpha$ implies that $\alpha = \alpha'$.

Without loss of generality, assume that $\deg(\alpha) \le \deg(\alpha')$ and that $\deg(\alpha)$ is of smallest degree among the $\{{}_i\alpha^j_{t-1}\}$ for which there is an α' with $\alpha \le \alpha' \le \alpha$ and $\alpha \ne \alpha'$. Moreover there is a shortest sequence

$\{\alpha_i\}_{i=0,\ldots,n} \subset \{_i\alpha^j_{t-1}\}$ of distinct simplices for $i \neq 0,n$ such that $\alpha = \alpha_0 = \alpha_n$, for some $0 < m < n$, $\alpha_m = \alpha'$, and $\alpha_i \circled{\leq} \alpha_{i+1}$ for $i = 0,\ldots,n-1$. Note that $\deg(\alpha) \leq \deg(\alpha_i)$ by the choice of an α of minimal degree.

$\mathrm{Deg}(\alpha)$ even is impossible since $p \circ F|(I \times \alpha_0 \cup I \times \alpha_1)$ have a non-empty singularity set with $\deg(\alpha_0) \leq \deg(\alpha_1)$. Thus $\deg(\alpha_0)$ is odd. Furthermore any singularities arising in $p \circ F|(I \times \alpha_0 \cup I \times \alpha_i)$, $i \neq n$, are all contained in the intersection of a single simplex of the original singularity set with $I \times \alpha_0$.

Suppose that α_i is the first in the sequence whose degree is not equal to that of α_0. By construction the singularity which provides the comparison between α_i and α_{i-1} is completely covered by the image from α_i. Furthermore since $\deg(\alpha_i) < \deg(\alpha_{i-1})$ the subdivision sequence implies that this singularity completely covers all simplices, of smaller degree, which can be compared to α_{i-1}. Thus if $i \geq 2$ the sequence can be shortened to $\alpha_0 \leq \alpha_i \cdots \leq \alpha_n = \alpha$ since α_0 and α_i are comparable. As this was the shortest sequence, either $i = 1$ or all degrees are equal. If $i \geq 1$, α_i is comparable to α_{n-1} so that either $\alpha_1 \leq \alpha_{n-1}$ or $\alpha_{n-1} \leq \alpha_1$. In the latter case the sequence can be shortened to $\alpha_1 \leq \cdots \leq \alpha_{n-1}\, \alpha_1$, which is impossible. In the former case the resulting sequence is $\alpha_0 \leq \alpha_1 \leq \alpha_{n-1} \leq \alpha_n$. This is also impossible as this implies that the levels of the simplices providing the comparison have this order.

Thus it must be that $\deg(\alpha_i) = \deg(\alpha_0)$, $i = 1,\ldots,n$. In this case all the simplices providing the comparisons arise from simplices in the original singularity set have the same image under $p \circ F$. This implies that their images under F have the ordering indicated by the sequence. This, as above, is clearly impossible. As a consequence there is no shortest sequence, and hence no such sequences at all, so that $\leq$ is a partial order on $\{{}_i\alpha^j_{t-2}\}$.

Since α's of different dimension are not comparable via $\leq$, it is easily seen that $\leq$ is contained in a well ordering of $\{{}_i\alpha^j_{t-2}\}$ such that $\alpha \leq \alpha'$ implies that $\dim\alpha \leq \dim\alpha'$. Since there are only finitely many α's this well ordering can be indexed, e.g., $\alpha_0 < \alpha_1 < \cdots < \alpha_t$. The sequence of fibered cylindrical collapses taking $I \times \alpha_i$ to $(I \times (\partial\alpha_i)) \cup (\{0\} \times \alpha_i)$ in order of decreasing index provides a fibered cylindrical sunny collapse of $F(J)$ to $F(J_o \cup (J \cap \{0\} \times K \times \Delta^s) \cup S_t)$. The "sunny" conclusion follows immediately from the partial order since $\alpha \leq \alpha'$ implies that no portion of $F(I \times (\operatorname{int}\alpha))$ lies above $F(I \times (\operatorname{int}\alpha'))$. Finally the dimensional assumption implies that the simplices of which α is a face occur later in the sequence and hence are collapsed previous to the α collapse.

SECTION 5. FIBERED LOCAL MODIFICATIONS VIA ENGULFING

In Hudson's proof that "concordance implies isotopy" the basic ingredient is a local modification technique, cf. Lemma 9.5 [38] or Proposition 5.1 [39]. In this a general position result is used to achieve the description of a sunny collapsing which, in turn, provides a prescription for the sequence of modifications. These arise from Zeeman's unknotting of balls in balls.

The requirements of the fibered situation are such that the analogous argument is not entirely possible. First the sunny collapsing is not terminal, leading to a dimensional restriction, and, second, the unknotting approach seems to lead to a codimension restriction. It appears that the first restriction is essential. The second is avoided by employing an engulfing argument in place of unknotting. As this new dimensional restriction is trivially satisfied in Hudson's case, a new proof of his "concordance implies isotopy" for manifolds is included.

5.1 THEOREM. Let $f\colon D^m \longrightarrow N$ be a proper locally unknotted piecewise linear embedding and

$$\begin{array}{ccc} F\colon I \times (D^m, \partial D^m) \times \Delta^s & \longrightarrow & I \times (N, \partial N) \times \Delta^s \\ \Big\downarrow p(3) & & \Big\downarrow p(3) \\ 1\colon \Delta^s & \longrightarrow & \Delta^s \end{array}$$

be a proper fibered piecewise linear locally unknotted embedding such that F restricts to $1 \times f \times 1$ on $(\{0\} \times D^m \times \Delta^s) \cup (I \times \partial D^m \times \Delta^s) \cup (I \times D^m \times \partial\Delta^s)$. If $\Pi_j(N) = 0$, $j \le k$, $n - m \ge 3$, $s \le n + k - m - 3$, and U is an open neighborhood of $f(D^m)$ in N there is a proper fibered piecewise linear ambient isotopy of $I \times (N, \partial N) \times \Delta^s$, H_t, which is fixed on the set

$B = (\{0\} \times N \times \Delta^s) \cup (I \times \partial N \times \Delta^s) \cup (I \times N \times \partial\Delta^s)$, and has the property that $H_1 \circ F(I \times D^m \times \Delta^s) \subset I \times U \times \Delta^s$.

Before beginning the proof of Theorem 5.1 the fibered engulfing lemma will be given. A proof can be found in [52] (Lemma 4.7), as well as several other places in the literature, and is a direct generalization of Zeeman's engulfing method.

5.2 LEMMA. Let $K_o \subset K_1$ be subcomplexes of $N \times \Delta^s$ such that $K_1 \searrow^{es} K_o$ is an elementary fibered simplicial collapse (cf. Definition 4.2) with $K_1 \setminus K_o \subset \mathrm{int}(N \times \Delta^s)$. Given $K_o \subset U$ and $(K_1 \setminus K_o) \subset V$, neighborhoods, there is a proper fibered piecewise linear ambient isotopy of $(N,\partial N) \times \Delta^s$, H_t, which is fixed outside V, is such that $K_o \subset H_t(U)$ for $0 \le t \le 1$, and is such that $K_1 \subset H_1(U)$.

A direct consequence of this lemma and the definition of a fibered simplicial collapse (cf. Definition 4.4) is the engulfing lemma required in the proof of Theorem 5.1.

5.3 LEMMA. Let $K_o \subset K_1$ be subcomplexes of $N \times \Delta^s$ such that $K_1 \searrow K_o$ is a fibered simplicial collapse with $K_1 \setminus K_o \subset \mathrm{int}(N \times \Delta^s)$. Given neighborhoods $K_o \subset U$ and $(K_1 \setminus K_o) \subset V$ there is a proper fibered piecewise linear ambient isotopy of $(N,\partial N) \times \Delta^s$, H_t, which is fixed outside V, is such that $K_o \subset H_t(U)$ for $0 \le t \le 1$, and $K_1 \subset H_1(U)$.

5.4 REMARK. Since, as noted previously, every fibered cylindrical collapse may be achieved by a fibered simplicial collapse this lemma also holds in the cylindrical case.

5.5 PROOF OF THEOREM 5.1. Let V be an open neighborhood of $f(D^m)$ whose closure is contained in U. Since $\overline{V}$ is compact there is a positive ϵ with the property that given any ϵ homeomorphism, g, of $I \times N \times \Delta^s$ then $g(I \times f(D^m) \times \Delta^s) \subset I \times V \times \Delta^s$ and $g^{-1}(I \times \overline{V} \times \Delta^s) \subset I \times U \times \Delta^s$.

Given such an ϵ, let φ_t be the proper fibered piecewise linear ambient isotopy whose existence is assured by Lemma 3.12. Let $I \times (D^m, \partial D^m) \times \Delta^s$ and $I \times (N, \partial N) \times \Delta^s$ be triangulated so that $\varphi_1 \circ F$: $I \times (D^m, \partial D^m) \times \Delta^s \to I \times (N, \partial N) \times \Delta^s$ and $p(3)$, on both range and domain of $\varphi_1 \circ F$, are simplicial. Furthermore require that every simplex of $\varphi_1 \circ F(I \times D^m \times \Delta^s)$ having a face in $(\{0\} \times N \times \Delta^s) \cup (I \times \partial N \times \Delta^s) \cup (I \times N \times \partial\Delta^s)$ lies entirely within $I \times V \times \Delta^s$. Note that this latter property remains in effect under any further subdivision.

As previously noted, in Remark 4.18, the singularity sets provided by the conclusion of Lemma 3.12 can be used to achieve a useful collapsing. Thus there is a fibered cylindrical sunny collapse of $\varphi_1 \circ F(I \times D^m \times \Delta^s)$ to $\varphi_1 \circ F((\{0\} \times D^m \times \Delta^s) \cup (I \times \partial D^m \times \Delta^s) \cup (I \times D^m \times \partial\Delta^s) \cup S_{r+1})$ in $I \times N \times \Delta^s$, by Corollary 4.17.

There are several observations and conventions which are useful in giving the proof of the theorem. First, if $\alpha_1, \ldots, \alpha_\ell$ denote the simplices employed in the description of the associated sequence of elementary fibered cylindrical sunny collapses then, from the proof of Lemma 4.16, these simplices may be ordered so that $i < j$ implies that $p(3)\dim(\alpha_i) \geq p(3)\dim(\alpha_j)$ and, if equal, then $\dim(\alpha_i) \geq \dim(\alpha_j)$. Let $\{K_i'\}_{0 \leq t \leq \ell}$ denote the successive stages of the sequence of elementary cylindrical sunny collapses from K_0' to K_ℓ'. Let $K_i = (\varphi_1 \circ F)^{-1}(K_i')$, employing the convention of Definition 3.11 for $\varphi_1 \circ F$.

5.6 A neighborhood, V, of K_i in $I \times D^m \times \Delta^s$ is <u>vertical</u> if there is a neighborhood U of $p(K_i \setminus (\{0\} \times D^m \times \Delta^s))$ in $D^m \times \Delta^s$ and an $\epsilon > 0$ such that $V = (I \times U) \cup ([0,\epsilon] \times D^m \times \Delta^s)$. It is easily seen that every neighborhood of K_i in $I \times D^m \times \Delta^s$ contains a vertical neighborhood.

5.7 A neighborhood, M, of K_i in $I \times D^m \times \Delta^s$ is <u>slanted</u> if the boundary of M in $I \times D^m \times \Delta^s$, ∂M, projects nonsingularly to $D^m \times \Delta^s$.

5.8 If L is any subcomplex of a complex $I \times K$, K arbitrary, let
$\overline{L} = \{(t,x) \mid \text{there is an } s \leq t \text{ such that } (s,x) \in L\}$ and
$\underline{L} = \{(t,x) \mid \text{there is an } s \geq t \text{ such that } (s,x) \in L\}$.

The first stage of the argument consists of engulfing $\overline{K'_\ell}$.

5.9 LEMMA. There is a proper fibered piecewise linear ambient isotopy, H_t, of $I \times (N,\partial N) \times \Delta^s$ which is fixed on B (as in the conclusion of the theorem) and such that $\overline{K'_\ell} \subset H_1(I \times V \times \Delta^s)$.

PROOF OF LEMMA 5.9. From Lemma 3.12 note that dimension $\mathcal{S}_{r+1} \leq s + m - n + 3$ so that dimension K'_ℓ, modulo $I \times V \times \Delta^s$ is no larger than $s + m - n + 3$. Since N is k-connected and, by assumption, $s + m - n + 3$ is no larger than k there is a proper fibered homotopy,
$h\colon I \times (p(2,3)K'_\ell, p(2,3)(K'_\ell \cap B)) \to (N \times \Delta^s, \partial(N \times \Delta^s))$ such that $h(\{1\} \times p(2,3)K'_\ell) \subset V \times \Delta^s$, it is fixed on $p(2,3)(K'_\ell \cap B)$, and it is in fibered general position modulo this fixed set.

Let A denote the singularities arising from an intersection with $h(\{0\} \times p(2,3)(K'_\ell))$. By Lemma 3.4, $\dim A \leq (s+m-n+3) + (s+m-n+4) - (s-n) \leq s + 2m - 3n + 7$. Since $n - m \geq 3$, $\dim \overline{A} \leq s + m - n + 2$. Thus an inductive argument can be given, in the standard manner, to imply that

$p(2,3)(\overline{A}')$ can be engulfed by $V \times \Delta^s$. This, however, is not quite sufficient. Since the projection of $S_1(h)$ into K_ℓ' can be shown to have local codimension at least one, a second singular set, $S_2(H)$, is defined via Lemma 3.4, and $\dim \overline{S_2(h)} \le s + m - n + 2$ so that $p(2,3)(\overline{S_2(h)'})$ can be engulfed by $V \times \Delta^s$, in addition to $p(2,3)(\overline{A}')$.

The fibered cylindrical collapse of $I \times K_\ell'$ to $I \times (K_\ell' \cap B') \cup \overline{A}' \cup \overline{S_2(h)}'$ has the property that the projection of each cell to $N \times \Delta^s$ is non singular. This series of collapses can be employed to give, via Lemma 5.3 and Remark 5.4, a complete engulfing of $p(2,3)(K_\ell')$ keeping $\partial(N \times \Delta^s)$ fixed. Note that the track of h is not covered in this last stage and, fortunately, this is not necessary. Let $G_t\colon I \times (N \times \Delta^s) \longrightarrow I \times (N \times \Delta^s)$ denote this engulfing. Since $K_\ell' \cap B \subset I \times V \times \Delta^s$, G_t can be employed to define the required engulfing by selecting a function $\theta\colon I \to I$, such that $\theta(0) = 0$ and $p(2,3)(K_\ell' \cap p(1)^{-1}(\sup(1-\theta)))$ is contained in $G_t(V \times \Delta^s)$, $0 \le t \le 1$, and defining $H_t(s,x,y) = (s, G_{t\theta(t)}(x,y))$. H_t is fixed on B since G_t is fixed on $\partial(N \times \Delta^s)$ and $\theta(0) = 0$. $\overline{K_\ell'} \subset H_1^{-1}(I \times V \times \Delta^s)$ since $p(2,3)(K_\ell') \subset G_1(V \times \Delta^s)$ and $p(2,3)(K_\ell' \cap p(1)^{-1}(\sup(1-\theta))) \subset G_t(V \times \Delta^s)$, $0 \le t \le 1$. This proves the lemma.

This provides the engulfing necessary for the application of the following lemma whose proof concludes the proof of the theorem.

5.10(i) STATEMENT. There is a neighborhood, V_i, of K_i in K_o containing the closure of a slanted neighborhood W_i of K_i in K_o and a proper fibered piecewise linear ambient isotopy, H_t^i, of $I \times (N,\partial N) \times \Delta^s$ which is fixed on B and has the property that $(V_i' \cup \overline{(V' \backslash W')}) \subset H_1^{-1}(I \times V \times \Delta^s)$.

5.11 LEMMA. Statement 5.10(i) implies Statement 5.10(i-1).

REMARK. Lemma 5.9 implies 5.10(ℓ) by taking $H_t^{\ell} = H_t$, $V'_{\ell} = K'_o \cap H_1(I \times V \times \Delta^s)$, and W_{ℓ} equal to any slanted neighborhood of K_{ℓ} in K_o whose closure lies in V_{ℓ}. The conclusion of the statement follows easily since $\overline{V'_{\ell}} \subset H_1(I \times V \times \Delta^s)$. Thus Lemma 5.11 provides the necessary inductive argument to complete the proof of the theorem.

5.12 PROOF OF LEMMA 5.11. Suppose that H_t^i is given engulfing $(V'_i \cup \overline{(V'_i \setminus W'_i)})$. Since $K_{i-1} \searrow K_i$ is an elementary cylindrical sunny collapse $\mathrm{sh}K'_i \cap (I \times \alpha_i)' = (\{0\} \times \alpha_i \cup I \times \partial\alpha_i)'$ and therefore $\overline{((I \times \alpha'_i)\setminus W'_i)} \cap K'_i = \emptyset$. Furthermore, since $K' \searrow K'_i$ is a cylindrical collapse, $K'_i \cap (I \times \alpha_i)' \subset K'_i \subset W_i$. Therefore there is an open neighborhood, $\mathcal{O}$ of $\overline{((I \times \alpha'_i)\setminus W'_i)}$ in $I \times N \times \Delta^s$ such that $\mathcal{O} = \overleftarrow{\mathcal{O}}$, $\underline{\mathcal{O}} \cap K \subset (W_i \cup (I \times \alpha_i))'$ and $\mathcal{O} \cap \partial(I \times N \times \Delta^s) \subset \{I\} \times (\mathrm{int}\ N) \times \Delta^s$.

Consider "K_1" $= K'_{i-1} \cup \overline{(I \times \alpha_i)'}$, "$K_o$" $= K'_i \cup \overline{((I \times \alpha_i)' \setminus W'_i)}$, "$V$" $= \mathcal{O}$, and "U" $= H_1^i(I \times V \times \Delta^s)$ in the terminology of Lemma 5.3. Note that although $K_1 \setminus K_o \subset V$ and $K_o \subset U$ a direct application of Lemma 5.3 is not possible since $K_1 \setminus K_o \not\subset \mathrm{int}(I \times N \times \Delta^s)$ and it is not clear that there is a fibered simplicial collapse $K_1 \searrow K_o$. These difficulties are not essential and are handled in two stages.

First note that $\mathrm{cl}(K_1 \setminus K_o) = \overline{((I \times \alpha_i) \setminus W_i)'} = L$. There is a "vertical" fibered simplicial collapse in $I \times N \times \Delta^s$ of L to $\overline{((I \times \alpha_i) \cap \partial W_i)'} \cup (L \cap \{1\} \times N \times \Delta^s)$, cf. figures 5.13 and 5.14, by making the vertical projection simplicial and employing a vertical cylindrical collapse. Since $p(2,3)$ is nonsingular on $(I \times \alpha_i)'$, $L \cap \{1\} \times N \times \Delta^s$ admits a fibered simplicial collapse to $\overline{((I \times \alpha_i) \cap \partial W_i)'}$.

The latter collapse takes place in $\{1\} \times N \times \Delta^s$ so that Lemma 5.3 can be applied to $\{1\} \times N \times \Delta^s$ to construct a fibered isotopy, whose support is

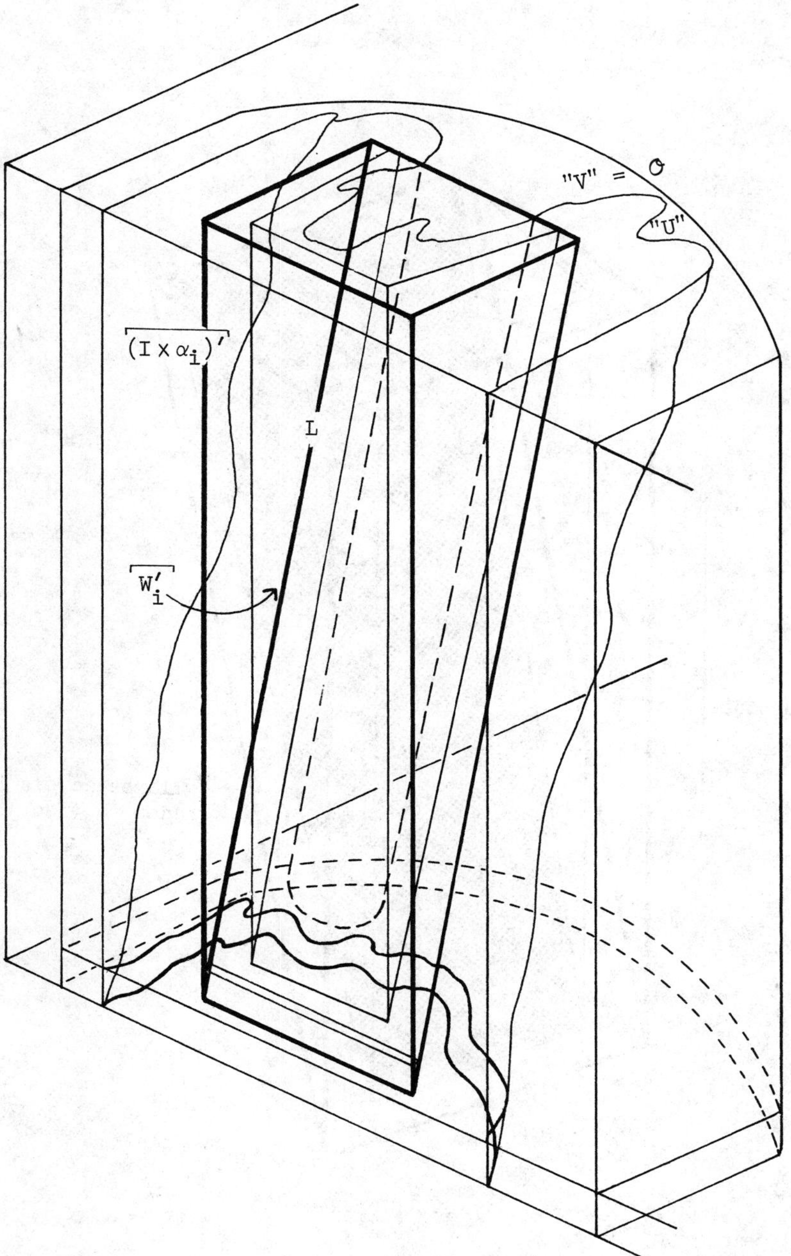

FIGURE 5.13

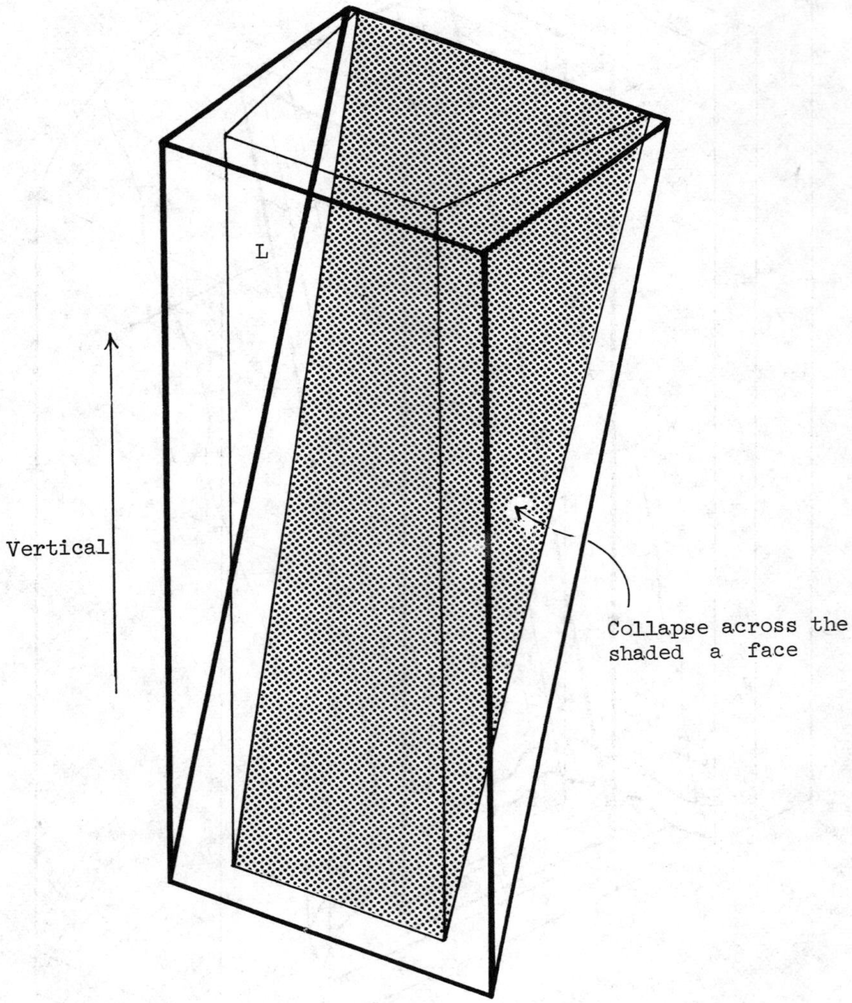

FIGURE 5.14

in $\mathcal{O} \cap \{1\} \times N \times \Delta^s$, to engulf this collapse. Since, in the generic notation, $K_o \subset H_t(V)$ for $0 \le t \le 1$ this engulfing can be extended to $\tilde{H}_t$ on $I \times N \times \Delta^s$, with support in $\mathcal{O}$, so as to achieve the conditions necessary to apply Lemma 5.3 to the former collapse with $K_1 = K'_{i-1} \cup \overline{(I \times \alpha_i)'}$, $K_o = K'_i \cup \overline{((I \times \alpha_i)' \setminus W'_i)} \cup (L \cap \{1\} \times N \times \Delta^s)$, $V = \mathcal{O}$ and $U = \tilde{H}_1 H^i_1(I \times V \times \Delta^s)$, and construct another engulfing, $\hat{H}_t$ such that $\sup \hat{H}_t \tilde{H} \subset \mathcal{O}$ and $K'_i \cup \overline{(I \times \alpha_i)'} \subset \hat{H}_1 \tilde{H}_1 H^i_1(I \times V \times \Delta^s)$.

Since $K'_{i-1} \subset K'_i \cup \overline{(I \times \alpha_i)'}$ all that remains is the construction of V_{i-1} and W_{i-1} as in statement 5.10(i-1) with $H^{i-1}_t = \hat{H}_t \tilde{H}_t H^i_t$. As noted above, $\overline{(I \times \alpha_i)'} \subset H^{i-1}_t(I \times V \times \Delta^s)$ so that the same is true for $I \times \tilde{W}$, a neighborhood of $I \times \alpha_i$ in K. That is $\overline{(I \times \tilde{W})'} \subset H^{i-1}_1(I \times V \times \Delta^s)$ and hence if $V'_{i-1} = (V'_i \setminus \mathcal{O}) \cup (I \times \text{int } \tilde{W})'$ then $V'_{i-1} \subset H^{i-1}_1(I \times V \times \Delta^s)$ since the support of H^{i-1}_t is contained in $\mathcal{O}$.

Finally as W_i is a slanted neighborhood of K_i in K there is a non-zero piecewise linear function $\lambda\colon \tilde{W} \longrightarrow [0,1]$ such that $W_i \cap (I \times \tilde{W}) = \{(x,t) \mid 0 \le t \le \lambda(x)\}$. Since $\alpha_i \subset \text{int } \tilde{W}$ there is a piecewise linear function $\delta\colon \tilde{W} \longrightarrow [0,1]$ such that the support of $\delta - \lambda$ is in the interior of $\tilde{W}$ and $I \times \delta^{-1}(1)$ is a neighborhood of $I \times \alpha_i$. As a consequence $W_{i-1} = W_i \setminus (I \times \text{int } \tilde{W}) \cup \{(x,t) \mid x \in \tilde{W},\ 0 \le t \le \delta(x)\}$ is a slanted neighborhood of K_{i-1} whose closure lies in V_{i-1}.

Consider $x \in \overline{(V_{i-1} \setminus W'_{i-1})}$. If $x \notin \mathcal{O}$ then either $x \in \overline{I \times \text{int } \tilde{W}}$, in which case it is in $H^{i-1}_1(I \times V \times \Delta^s)$, or else it is in $\overline{V'_i \setminus W'_i}$ so that it is also in $H^{i-1}_1(I \times V \times \Delta^s)$ as support $\hat{H}_t \tilde{H}_t \subset \mathcal{O}$. If, however $x \in \mathcal{O}$ then $x \in \overline{(I \times \text{int } \tilde{W})'}$ in which case it is in $H^{i-1}_1(I \times V \times \Delta^s)$. Therefore

$$(V'_{i-1} \cup \overline{(V'_{i-1} \setminus W'_{i-1})}) \subset H^{i-1}_1(I \times V \times \Delta^s)$$

concluding the proof of Theorem 5.1.

SECTION 6. FIBERED GLOBAL MODIFICATIONS

In this, the final section of the paper, is found the proof of Theorem 1.8, on which the proofs of the major results of the paper depend, as indicated in §1.

THEOREM 1.8. Let f be a proper piecewise linear embedding, $f\colon (D^m,\partial D^m) \to (N,\partial N)$, $\Pi_j(N) = 0$, $j \le k$. If $n - m \ge 3$, $s \le n + k - m - 3$, and

$$F\colon I \times (D^m,\partial D^m) \times \Delta^s \longrightarrow I \times (N,\partial N) \times \Delta^s$$

is a proper fibered piecewise linear embedding such that

$$F \mid ((I \times \partial D^m \times \Delta^s) \cup (I \times D^m \times \partial\Delta^s)) = 1 \times f \times 1,$$

then there is a proper fibered piecewise linear ambient isotopy H_t of $I \times (N,\partial N) \times \Delta^s$, fixed on $(\{0\} \times N \times \Delta^s) \cup (I \times \partial N \times \Delta^s) \cup (I \times N \times \partial\Delta^s)$ such that $H_1 \circ F = (1, F \mid (\{0\} \times D^m \times \Delta^s))$.

6.1 PROOF OF THEOREM 1.8. There is a proper fibered piecewise linear ambient isotopy $H_t'\colon \{0\} \times (N,\partial N) \times \Delta^s \to \{0\} \times (N,\partial N) \times \Delta$ such that $H_1' \circ F \mid (\{0\} \times (D^m,\partial D^m) \times \Delta^s) = 1 \times f \times 1$ by the fibered isotopy extension theorem of Hudson [37]. Let $F' = (1,H_1') \circ F\colon I \times (D^m,\partial D^m) \times \Delta^s \to I \times (N,\partial) \times \Delta^s$ and note that

$$F' \mid ((\{0\} \times D^m \times \Delta^s) \cup (I \times \partial D^m \times \Delta^s) \cup (I \times D^m \times \partial\Delta^s)) = 1 \times f \times 1$$

since $F \mid ((I \times \partial D^m \times \Delta^s) \cup (I \times D^m \times \Delta^s)) = 1 \times f \times 1.$

Let B^n be a regular neighborhood of $f(D^m)$ in N and note that B^n is a piecewise linear n ball with $f\colon (D^m,\partial D^m) \to (B^n,\partial B^n)$ a proper piecewise linear embedding [38]. Thus $1 \times f \times 1\colon I \times D^m \times \Delta^s \to I \times B^n \times \Delta^s$ is a proper fibered piecewise linear embedding of an $m+1$ ball into an $n+1$ ball.

By Theorem 5.1 there is a proper fibered piecewise linear ambient isotopy H''_t fixed on $((\{0\} \times N \times \Delta^s) \cup (I \times \partial N \times \Delta^s) \cup (I \times N \times \Delta^s))$ such that $H''_1 \circ F': I \times D^m \times \Delta^s \to I \times B^n \times \Delta^s$ is a proper fibered embedding of an $m+1$ ball into an $n+1$ ball with agrees with $1 \times f \times 1$ on $((\{0\} \times D^m \times \Delta^s) \cup (I \times \partial D^m \times \Delta^s) \cup (I \times D^m \times \partial\Delta^s))$ by taking U, in Theorem 5.7, to be $\mathrm{int}_N B^n$.

Applying the Alexander isotopy argument, there is a proper fibered piecewise linear ambient isotopy, H'''_t, of $\{1\} \times B^n \times \Delta^s$ fixed on $(\{1\} \times \partial B^n \times \Delta^s) \cup (\{1\} \times B^n \times \partial\Delta^s)$ such that $H'''_1 \circ H''_1 \circ F' \mid (\{1\} \times D^m \times \Delta^s) = 1 \times f \times 1$. H'''_t extends to a proper fibered piecewise linear ambient isotopy of $I \times B^n \times \Delta^s$ fixed on $(\{0\} \times B^n \times \Delta^s) \cup (I \times \partial B^n \times \Delta^s) \cup (I \times B^n \times \partial\Delta^s)$ by $H''_t(s,x) = (s, H'''_{\mu(s,t)}(x))$, where $\mu: I \times I \to I$ is a piecewise linear function such that $\mu(0,t) = \mu(s,0) = 0$, $\mu(1,t) = t$, and $\mu(s,1) = s$. Note that $H'''_1 \circ H'' \circ F' \mid (\{1\} \times D^m \times \Delta^s) = 1 \times f \times 1$.

Thus $H'''_1 \circ H''_1 \circ F'$ is a proper fibered piecewise linear embedding of $I \times D^m \times \Delta^s$ in $I \times B^n \times \Delta^s$ which agrees with $1 \times f \times 1$ on $\partial(I \times D^m \times \Delta^s)$. Thus, by the Alexander isotopy, there is a proper fibered piecewise linear ambient isotopy, H''''_t, of $I \times B^n \times \Delta^s$ fixed on $\partial(I \times B^n \times \Delta^s)$ such that $H''''_1 \circ H'''_1 \circ H''_1 \circ F' = 1 \times f \times 1$. Since both H'''_t and H''''_t extend to proper fibered ambient isotopies of $I \times N \times \Delta^s$ which are fixed on $(\{0\} \times N \times \Delta^s) \cup (I \times \partial N \times \Delta^s) \cup (I \times N \times \partial\Delta^s)$ by the identity on the complement of $I \times B^n \times \Delta^s$, also denoted by H'''_t and H''''_t, the desired isotopy of $I \times N \times \Delta^s$ can be defined by $H_t = (1, H'_1)^{-1} \circ H''''_t \circ H'''_t \circ H''_t \circ (1, H'_1)$.

H_t is fixed on $(\{0\} \times N \times \Delta^s) \cup (I \times \partial N \times \Delta^s) \cup (I \times N \times \partial\Delta^s)$ since each of H_t'', H_t''', and H_t'''' are fixed there. Finally, since $H_1'''' \circ H_1''' \circ H_1'' \circ (1,H_1')\ F = 1 \times f \times 1$ and $(1,H_1')^{-1}\ (1 \times f \times 1) = (1,F \mid (\{0\} \times D^m \times \Delta^s))$, $H_1 \circ F = (1,F \mid (\{0\} \times D^m \times \Delta^s))$.

References

1. Akiba, T., On the homotopy type of PL_2, J. Fac. Sci., Univ. Tokyo, Sect. I 14 (1967), 197-204.

2. ______, Homotopy types of some PL complexes, Bull. Amer. Math. Soc. 77 (1971), 1060-1062.

3. ______, Filling in cross sections of piecewise linear bundles, mimeograph 1971.

4. ______, PL_3 is homotopy equivalent to O_3, mimeograph 1971.

5. Alexander, J. W., On the deformation of an n-cell, Proc. Nat. Acad. Sci. Wash. 9 (1923), 406-407.

6. Antonelli, P., Burghelea, D., and Kahn, P., The concordance-homotopy groups of geometric automorphism groups, Lec. Notes in Math. Springer-Verlag (1971).

7. ______, The non-finite homotopy of some diffeomorphism groups, Topology 11 (1972), 1-50.

8. Arens, R., A topology for spaces of transformations, Ann. of Math. 2 (1952), 11-22.

9. Arens, R. and Dugundji, J., Topologies for function spaces, Pacific J. Math. 1 (1951), 5-31.

10. Barratt, M., Gugenheim, V., and Moore, J., On semisimplicial fibre-bundles, Amer. J. Math. 81 (1959), 639-651.

11. Berkowitz, H. and Dancis, J., PL approximations of embeddings and isotopies of manifolds in the 3/4 range, Topology of Manifolds, Markham (1970).

12. Browder, W., Open and closed disk bundles, Ann. of Math. 83 (1966), 218-230.

13. Brown, M., On a theorem of Fisher concerning the homeomorphism group of a manifold, Mich. Math. J. 9 (1962), 403-405.

14. Brown, M. and Gluck, H., Stable structures on manifolds: I, II, III, Ann. of Math. 79 (1964), 1-58.

15. Cerf, J., Topologie de certaines espaces des plongements, Bull. Sci. Math. France 89 (1961), 227-380.

16. ______, Isotopie et pseudo-isotopie, Proc. ICM, Moscow (1966), 429-437.

17. ______, Sur les diffeomorphismes de la sphere de dimension trois ($\Gamma_4 = 0$), Lect. Notes in Math. Vol. 53 (1968).

18. Cernavskii, A., Local contractibility of the homeomorphism group of a manifold, Soviet Math. Dokl. 9 (1968), 1171-1174.

19. Chenciner, A., Pseudo-isotopies differentiables et pseudo-isotopies lineaires par morceaux, Comptes Rendus de l'Acad. Sci. Paris 270 (1970), 1312-1315.

20. Connell, E., Approximating stable homeomorphisms by piecewise linear ones, Ann. of Math. 78 (1963), 326-338.

21. Curtis, E., Simplicial homotopy theory, Advances in Mathematics 6 (1971), 107-209.

22. Dancis, J., An introduction to embedding spaces for geometric topologists, Topology of Manifolds, Markham (1970).

23. Edwards, R. and Kirby, R., Deformations of spaces of embeddings, Ann. of Math. 93 (1971), 63-88.

24. Edwards, R. and Miller, R., Local connectivity of spaces of embeddings, Notices, AMS 19 (April 1972), A-467.

25. Eells, J. and Earle, C., A fibre bundle description of Tiechmuller theory, J. Differential Geometry 3 (1969), 19-43.

26. Fisher, G., On the group of all homeomorphisms of a manifold, Trans. AMS 97 (1960), 193-212.

27. Frechet, M., Sur quelques points due calcul fonctionnel, Recondicenti di Palermo 22 (196), 1-74.

28. Fox, R, On topologies for function spaces, Bull. Amer. Math. Soc. 51 (1945), 429-432.

29. Geoghegan, R., On spaces of homeomorphisms, embeddings and functions - I, Topology 11 (1972), 159-178.

30. Gluck, H., Embeddings in the trivial range, Ann. of Math. 81 (1965), 195-210.

31. Haefliger, A. and Poenaru, V., La classification des immersions combinatoires, IHES 23 (1964), 75-91.

32. Haefliger, A. and Wall, C.T.C., Piecewise linear bundles in the stable range, Topology 4 (1965), 209-214.

33. Hamstrom, M., A note on homotopy in homeomorphism spaces, Illinois J. Math. 9 (1965), 602-607.

34. Hausdorff, F., Grundzüge der Mengenlehre, Leipzig (1914). Reprinted by Chelsea, New York (1949).

35. Hirsch, M., On tubular neighborhoods of piecewise linear and topological manifolds, Conference on the Topology of Manifolds, Prindle, Weber and Schmidt (1968).

36. Hudson, J.F.P., Concordance and isotopy of PL embeddings, Bull. Amer. Math. Soc. 72 (1966), 534-535.

37. ______, Extending piecewise linear isotopies, Proc. London Math. Soc. (3) 16 (1966), 651-668.

38. ______, Piecewise Linear Topology, Benjamin, New York, 1969.

39. ______, Concordance, isotopy and diffeotopy, Ann. of Math. 91 (1970), 425-448.

40. James, I., On the iterated suspension, Quart. J. Math. J. Oxford Ser. 5 (1954), 1-10.

41. Kato, M., Combinatorial prebundles, I and II, Asaka J. Math. 4 (1967), 289-303, 305-311.

42. Kirby, R. C. and Siebenmann, L., On the triangulation of manifolds and the Hauptvermutung, Bull. Amer. Math. Soc. 75 (1969), 742-749.

43. Kneser, H., Die deformationssatze der einfach zusammenhangenden flachen, Math. Z. 25 (1926), 362-72.

44. Kuiper, N.H. and Lashof, R., Microbundles and bundles; I and II, Invent. Math. 1 (1966), 1-17, 243-259.

45. Lashof, R., Problems from Seattle Conference, Ann. of Math. 81 (1965), 565-591.

46. Lashof, R. and Rothenberg, M., Microbundles and smoothing topology, 3 (1965), 357-388.

47. ______, Triangulation of manifolds I and II, Bull. Amer. Math. Soc. 75 (1969), 750-754, 755-758.

48. Leslie, J., On a differentiable structure for the group of diffeomorphisms, Topology 6 (1967), 263-271.

49. Lusk, E., Homotopy groups of spaces of PL embeddings, Notices, AMS 16 (Nov. 1969), 1067.

50. Mason, W., The space of all self-homeomorphisms of a two-cell which fix the boundary is an absolute retract (to appear).

51. ______, The space H(M) of homeomorphisms of a compact manifold onto itself is homeomorphic to H(M) minus any σ-compact set (to appear).

52. Millett, K. C., Non singular sections to Euclidean bundles, Trans. Amer. Math. Soc. 144 (1969), 127-143.

53. ______, Homotopy normal bundles for locally flat immersions and embeddings of topological manifolds, Illinois J. Math. 13 (1969), 564-572.

54. ______, Normal structures for locally flat embeddings, Proc. Amer. Math. Soc. 20 (1969), 580-584.

55. ______, Homotopy sequences of fibrations, Fund. Math. 79 (1973), 49-62.

56. Milnor, J., Microbundles I, Topology 3 (1964), Suppl. 1, 53-80.

57. Morlet, C., Microfibres et structures differentiables, Sem. Bourbaki (1963/4), Fasc 1, Exp. 263.

58. ______, Isotopie et pseudo-isotopie, C. R. Acad. Sci. Paris 266 (1968), 559-560.

59. ______, Plongements et Automorphismes des Variétés, Notes Cours Peccot, College de France, 1969.

60. ______, Isotopie et pseudo-isotopie, C. R. Acad. Sci. Paris 268 (1969), 1080-1082.

61. Palais, R., Homotopy theory of infinite dimensional manifolds, Topology 5 (1966), 1-16.

62. Robertson, S., Retracting diffeomorphisms of spheres, Proc. Amer. Math. Soc. 24 (1970), 57-59.

63. Rourke, C., Embedded handle theory, concordance and isotopy, Topology of Manifolds, Markham, 1970.

64. Rourke, E. P. and Sanderson, B. J., Block Bundles: I, II, III, Ann. of Math. 87 (1968), 1-28, 256-278, 431-483.

65. ______, An embedding without a normal microbundle, Invent. Math. 3 (1967), 293-299.

66. ______, Δ-sets I: Homotopy theory, Δ-sets II: Blockbundles and block fibrations, Quart. J. Math. 22 (1971), 321-338.

67. Scott, G., The space of homeomorphisms of a 2-manifold, Topology 9 (1970), 97-109.

68. ______, On the homotopy type of PL_3, preprint 1970.

69. Smale, S., Diffeomorphisms of S^2, Proc. Amer. Math. Soc. 10 (1959), 621-626.

70. Stern, R., Stability implies normal and disc bundles, Bull. Amer. Math. Soc. 79 (1973), 133-135.

71. Wall, C.T.C., Locally flat PL submanifolds with codimension two, Proc. Cambridge Philos. Soc. 63 (1967), 5-8.

72. Williamson, Robert E., Jr., Cobordism of combinatorial manifolds, Ann. of Math. 83 (1966), 1-33.

73. Zeeman, E.C., Seminaire á l'I.H.E.S., 1962-163 (mimeo).

74. _______, Unknotting combinatorial balls, Ann. of Math. 78 (1963), 501-526.

University of California
Santa Barbara, California 93106